Test Your MATH

Steve Ryan

Sterling Publishing Co., Inc. New York

Contents

Edited by Claire Bazinet

10 9 8 7 6 5 4 3 2

Published by Sterling Publishing Company, Inc.
387 Park Avenue South, New York, N.Y. 10016

Distributed in Canada by Sterling Publishing
℅ Canadian Manda Group, P.O. Box 920, Station U
Toronto, Ontario, Canada M8Z 5P9
Distributed in Great Britain and Europe by Cassell PLC
Villiers House, 41/47 Strand, London WC2N 5JE, England
Distributed in Australia by Capricorn Link (Australia) Pty Ltd.
P.O. Box 6651, Baulkham Hills, Business Centre, NSW 2153, Australia
Manufactured in the United States of America

Sterling ISBN 0-8069-0724-X

A Note to the Reader

Put on your thinking caps and prepare to explore a macrocosm of mathematical puzzles, posers, pastimes, and paradoxes. Mathematics isn't just quantum theory. It takes the form of such popular contests as tic-tac-toe and chess. Galileo once described mathematics as the alphabet in which God has created the universe.

Before you is a competitive arena for the mind in which the rewards are great self-satisfaction. Much of the fun and fascination of solving mathematical diversions is derived from applying the precise logic that restores order to the chaos of a problem. With a little creative cunning you can restore harmony to these mathematical mind benders.

I often think of my puzzles as a bridge between mathematics and art. Each puzzle is esthetically designed to capture the eye, flirt with the curiosity, and tempt your thought processes to follow a thread of reason on a meandering journey of mathematical delight. Some puzzles ask you to apply the basics: addition, subtraction, multiplication, and division; while others require more abstract mathematical maneuvers, manipulations, and meditation. There are magic squares, devious dissections, tessellation teasers, problems in topology, and much more.

Puzzles vary in degree of complexity, but all are fair and require no formal mathematical training. The difficulty of each puzzle is rated as a one, two, or three penciller. Although the one-pencil puzzlers are the easiest, they still require considerable flexing of your mental muscles. Three-pencil puzzlers are the toughest nuts to crack, demanding rigorous cerebral calisthenics to decipher or divine these gruelers.

If you're up to the challenge, and I'm sure you are, you're in for a terrific think test and hours of fun. For, whether you're a "Mathemagician" or a "Number Novice," all of these puzzles are guaranteed to intrigue.

So sharpen your pencils, and sharpen your wits. You may be a lot more mathwise than you think!

Steve Ryan

1

BEER BARREL POKER

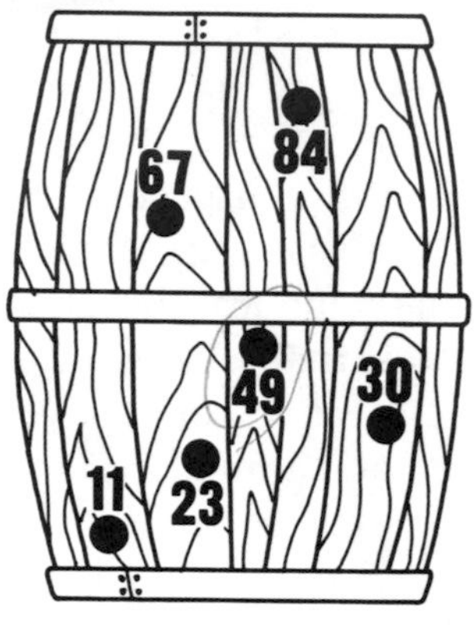

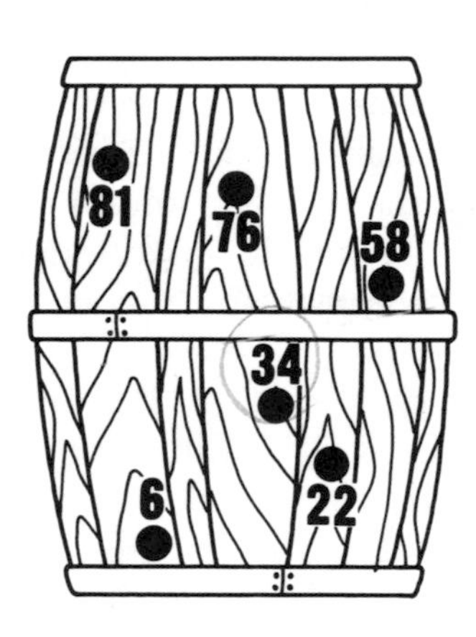

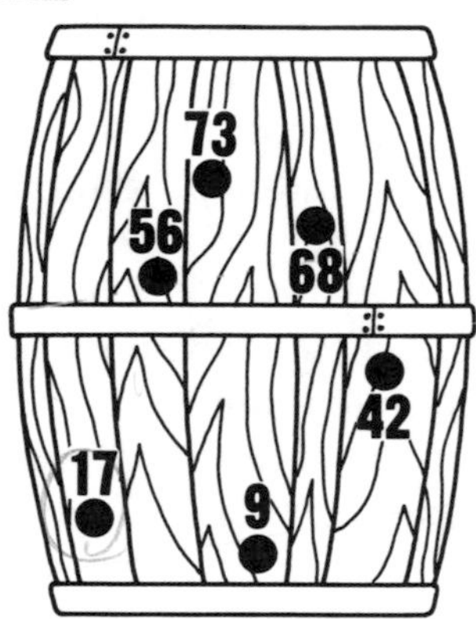

Each of these barrels contains 100 gallons of beer. Can you poke out a knothole in each barrel to leave a total of exactly 100 gallons in all three barrels combined? The number at each knothole shows the gallons that would remain in the given barrel.

Solution on page 82.

2

PUT IN SHAPE

Place the remaining numbers from one to ten in the seven divisions of this overlapping geometric configuration to fulfill the following requirements: 1) The circle, square, and triangle must individually total thirty. 2) The three outer divisions of the circle, square, and triangle must also total thirty.

Solution on page 84.

(Need a clue? Turn to page 10.)

3

IN TENTS

It is known that four officers are strategically located in four different tents that total thirty-two. Orders state that each horizontal, vertical, and diagonal row of four tents must quarter one officer. Which tents do the officers occupy?

Solution on page 86.

4

THE THIRD DEGREE

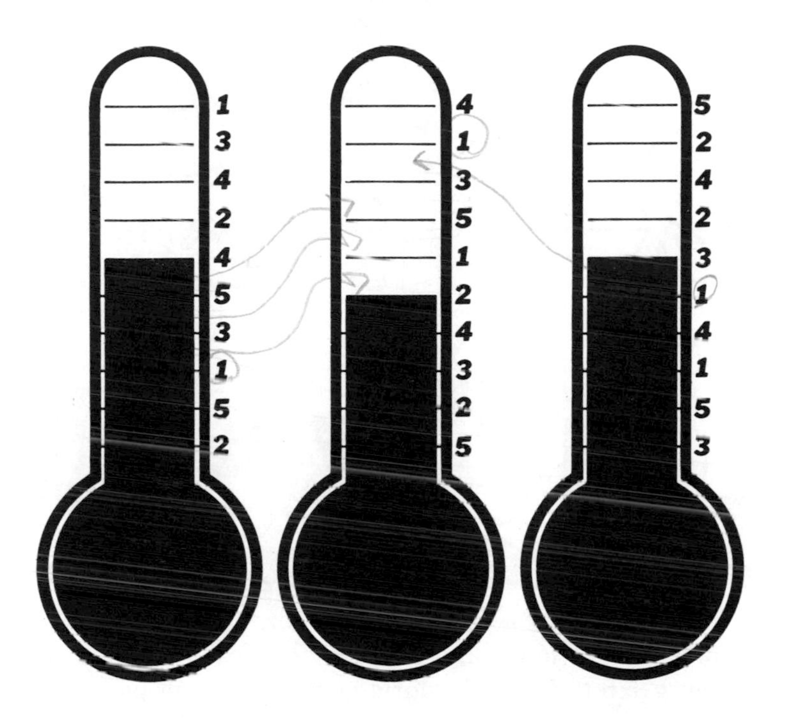

Your challenge is to balance the thermometers in this puzzle in such a way that they all read an identical number. For each unit rise in any thermometer, one of the other thermometers must fall one unit, and vice versa.

Solution on page 88.

5

CANDY CODED

It is known that each horizontal, vertical, and diagonal row of four candies totals 200 calories. You must determine the caloric content of each piece of candy from the following information:

Three candies have 20 calories apiece.
Two candies have 40 calories apiece.
Seven candies have 60 calories apiece.
Three candies have 80 calories apiece.
The black candy is calorie-free.

Solution on page 90.

6

SHADY HOLLOWS

There are eleven hollow shapes in this puzzle. It is your challenge to shade in four shapes which do not border on one another.

Solution on page 92.

7

MAD HATTER'S CAP SIZE

Here is a puzzle in which your challenge is to eliminate fractions. Add two or more of the cap sizes together to produce a whole number.

Solution on page 94.

(Need a clue? Turn to page 13.)

Clue to puzzle **2***: One division remains blank.*

8

TRUE TO ONE'S COLORS

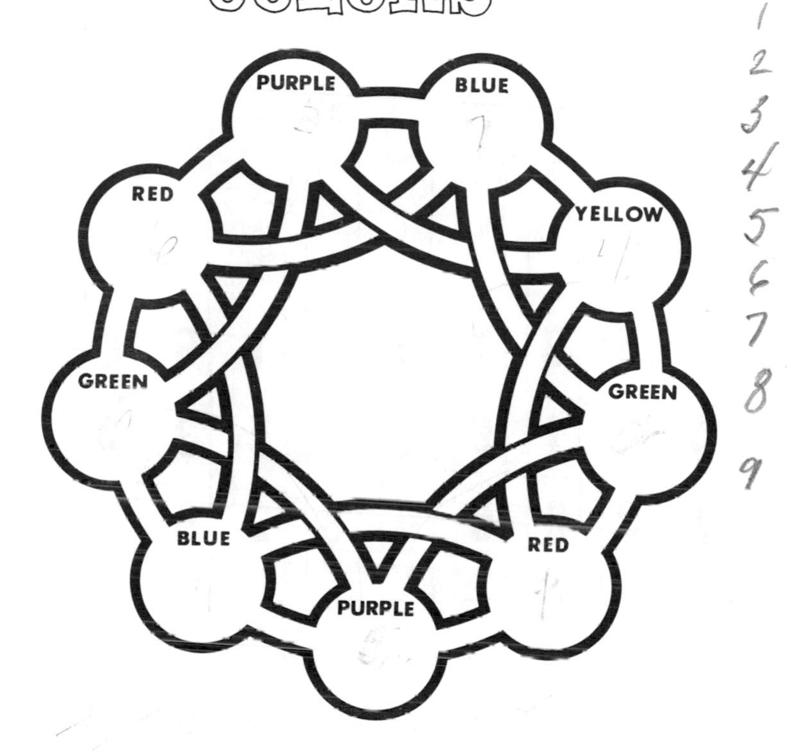

The interior lines of this puzzle crisscross but do not intersect. Place the numbers one through nine in the nine colored circles to fulfill the following requirements: 1) Any set of three numbers which totals fifteen (there are eight) must include three different colors. 2) Numbers of consecutive value may not be directly linked by any passage.

Solution on page 82.

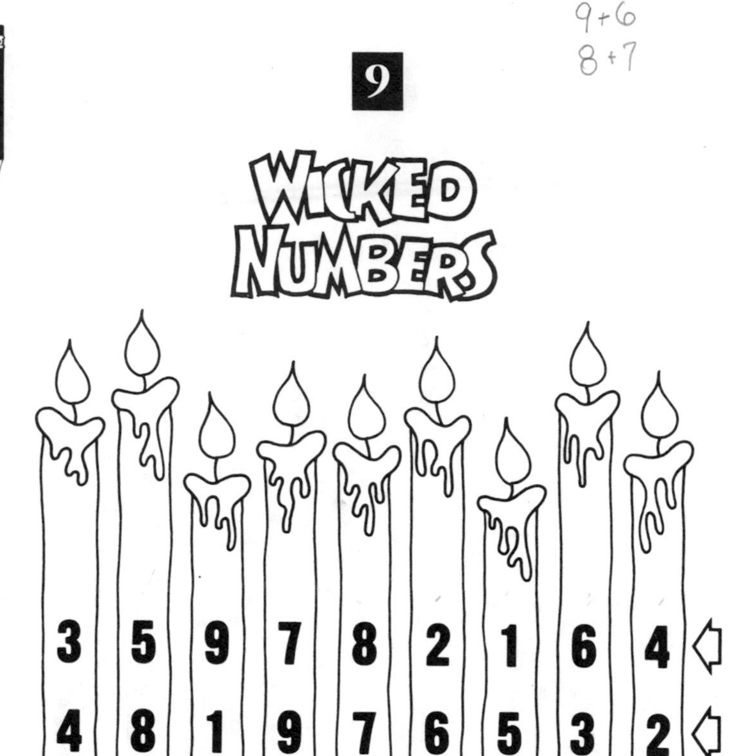

The numbers one through nine appear three times each in this puzzle. Your assignment is to blow out three candles which will total fifteen in each of the three horizontal rows. The three candles you select must carry the numbers one through nine. (No number may be used more than once.)

Solution on page 84.

(Need a clue? Turn to page 16.)

10

NIGHTWALKER

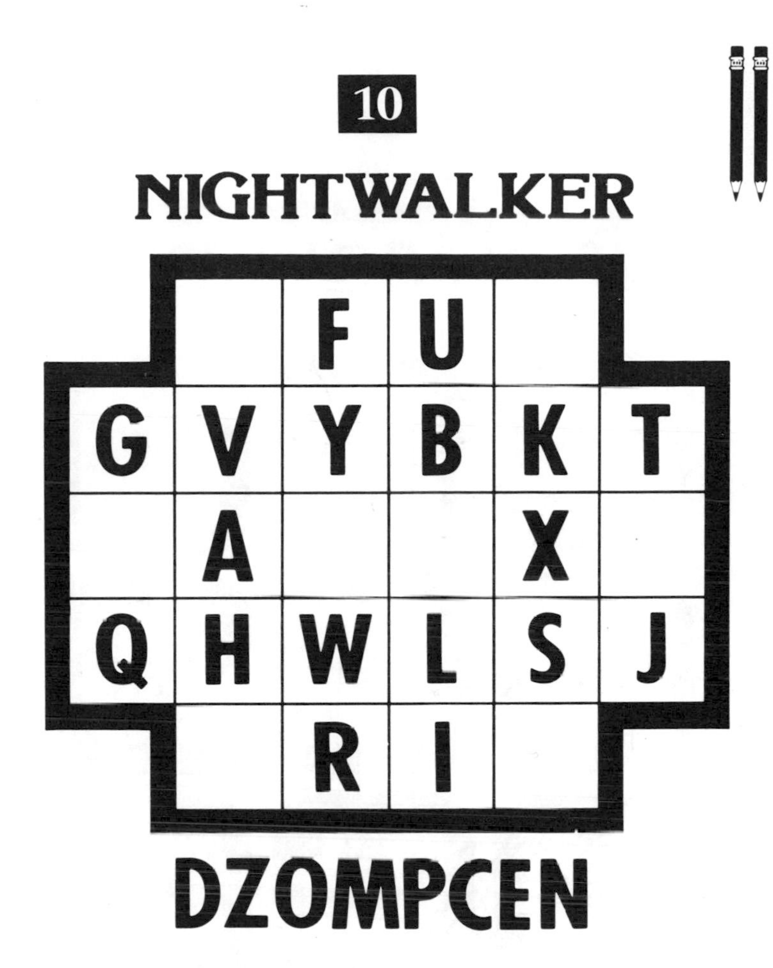

Position the eight remaining letters of the alphabet in the vacant squares of this puzzle to complete an alphabetical progression created by the moves of a chess knight.

Solution on page 86.

Clue to puzzle **7**: *Try "capsizing" one of the hats.*

11

MAGIC WORD SQUARES

APE	PEA	LEI
LAW	AWL	PRO
PLY	LEG	AIR

Each letter in this puzzle represents a different number from zero to nine. It is your challenge to switch these letters back to numbers in such a way that each horizontal, vertical, and diagonal row of three words totals the same number. Your total for this puzzle is 1446.

Solution on page 88.

(Need a clue? Turn to page 17.)

12

TIC-TAC-TOPOLOGY

Here's a strategy game of topology for two players. Simply force your opponent to connect three or more states with their Xs or Os and you win the game. Just as in tic-tac-toe, one player plays X and one player plays O. Players alternate positioning one of their marks per state until one player is forced to connect three or more states. In this sample game in progress, it is your move and challenge to position an O on the map in such a way that it will be impossible for your opponent to position an X without losing the game. Note: Only one X or O can be used to mark Michigan (a bridge is shown connecting both halves), but diagonally adjacent states, such as Arizona and Colorado, are not considered connected. You can enjoy playing this game with maps of other countries and continents.

Solution on page 90.

13

HANG BY A THREAD

Each of the numbers one through nine appears twice in the eighteen disks that are hanging by threads. Your task is to cut the least number of threads so as to drop one set of numbers and leave nine disks hanging that reveal the remaining set of numbers from one to nine.

Solution on page 92.

Clue to puzzle **9**: *Start by blowing out the middle candle.*

14

PIG STYMIE

Put nine pigs in eight pens.

Solution on page 94.

Clue to puzzle **11** *: APE = 473.*

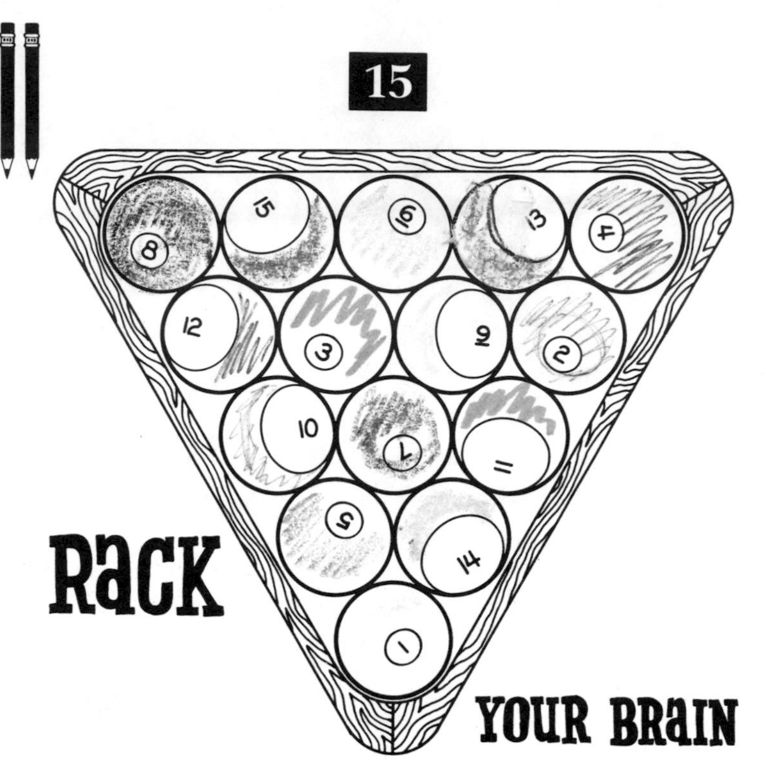

15

RACK YOUR BRAIN

An interesting game of pool involving three players has just been completed. It has a winner but the total point scores are as close as possible. Determine how the balls were distributed from the following information: 1) Each player has a different number of balls. 2) No player has two balls of consecutive number. 3) No player has two balls of identical color.

Ball colors: 1 and 9 are yellow, 2 and 10 are blue, 3 and 11 are red, 4 and 12 are purple, 5 and 13 are orange, 6 and 14 are green, 7 and 15 are maroon, 8 is black.

Solution on page 82.

(Need a clue? Turn to page 22.)

Make two straight cuts that divide this figure into four pieces of equal size and shape which, when rearranged, will form a square revealing another square within.

Solution on page 84.

(Need a clue? Turn to page 28.)

17

GEOMETRACTS

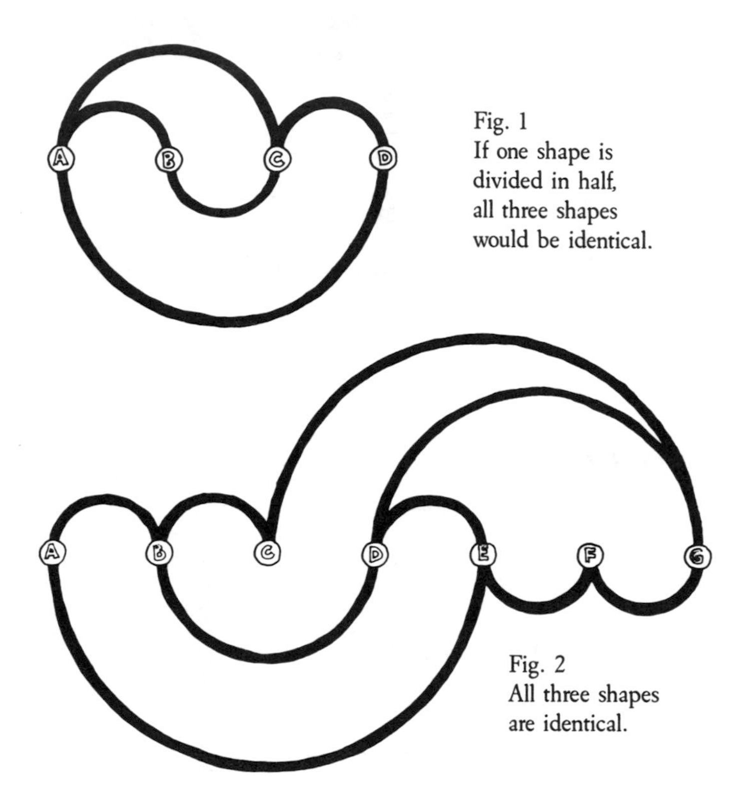

Here are two distorted geometric figures. Both have been stretched in such a way that the original figure is unrecognizable at first glance. Your task is to straighten all the lines in each figure to reveal its original identity. The circled letters designate the intersection of two or more lines. Vital clues are given for each figure.

Solution on page 86.

18

TIC-TAC-TOTAL

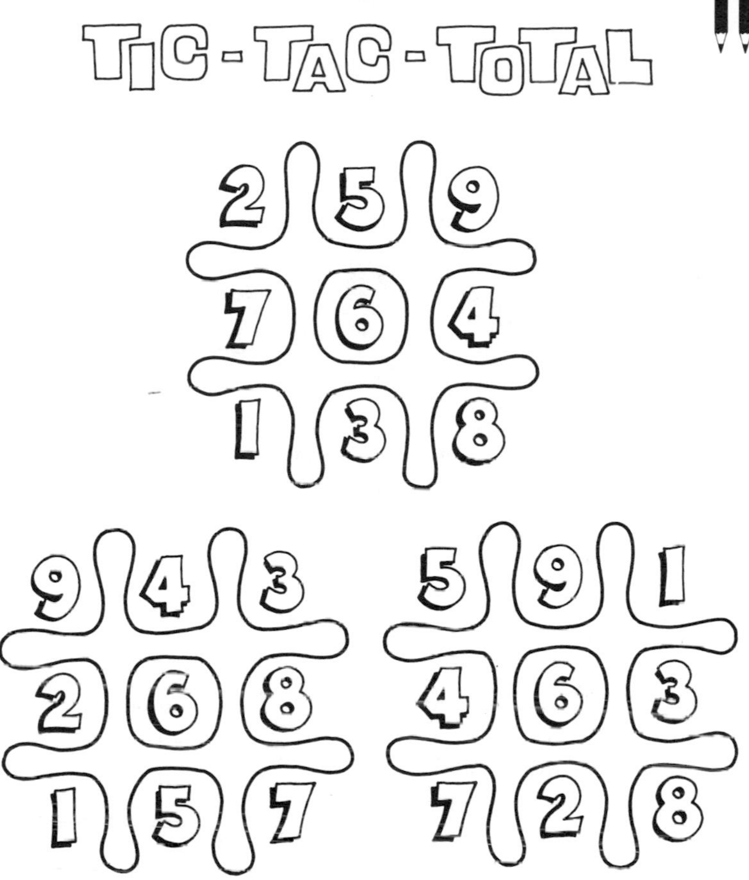

Your challenge in this puzzle is to circle a winning tic-tac-toe on each on the three game boards in the following manner: 1) One game must contain a diagonal win, one game must contain a horizontal win, and one game must contain a vertical win. 2) All numbers from one through nine must be circled in constructing these three winning lines.

Solution on page 88.

19

MENTAL BLOCKS

At present, the numbers 49,067 and 58,132 appear in these ten mental blocks. Can you switch the positions of any two blocks to create two new numbers so that one number will be twice as large as the other?

Solution on page 90.

Clue to puzzle **15**: *The player with the highest score has the least number of balls.*

The sorcerer's symbols are the lightning bolt, the crescent moon, and the star. Each symbol has a specific amount of magical power. The saucers in this puzzle reveal the magnitude of those powers. It is known that the three crisscrossing arrows point to saucers of equivalent powers. Your task is to determine which symbol or symbols, equalling the star, must be positioned in the empty saucer.

Solution on page 92.

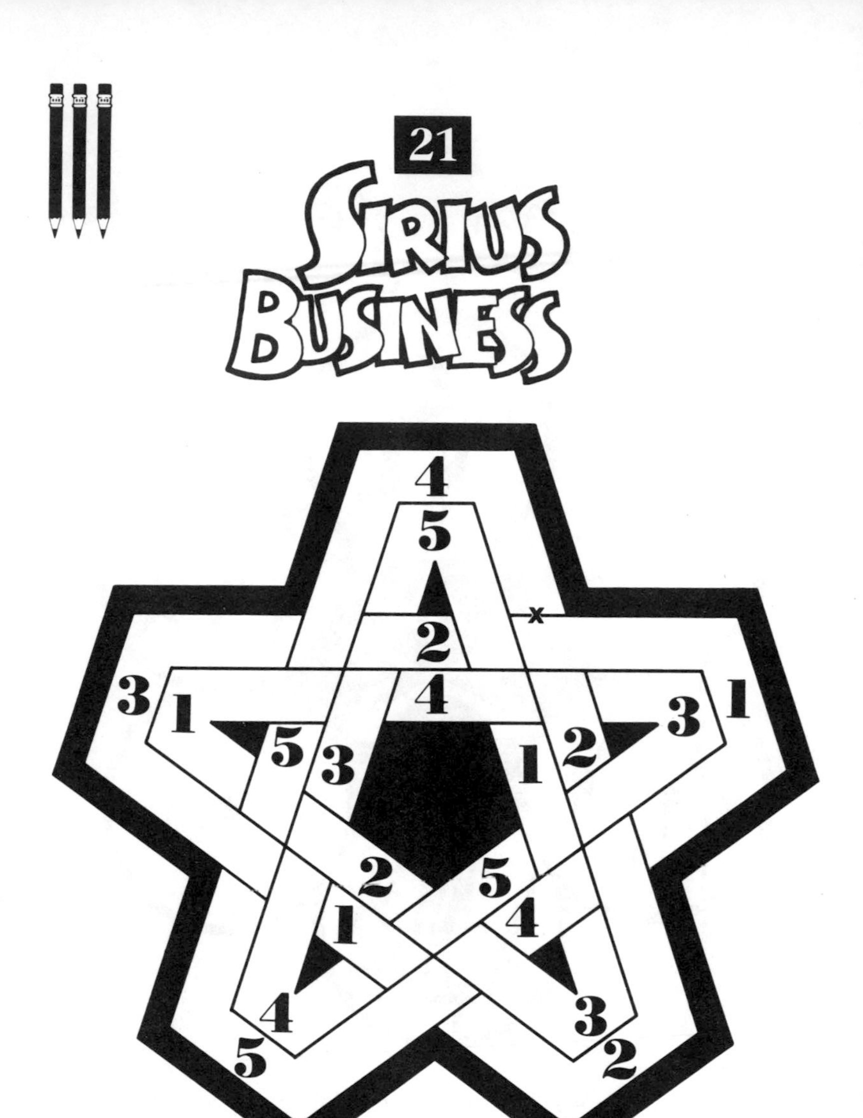

Cross out or remove sixteen of the short divider lines of this puzzle (one is given) in such a way that four large shapes, each containing the numbers one through five, are formed.

Solution on page 94.

22

COMMON CENTS

1¢	1 PENNY 1 NICKEL
1¢	2 PENNIES
1¢	2 PENNIES
1¢	2 PENNIES
1¢	1 PENNY 1 NICKEL
5¢	1 PENNY 1 NICKEL 2 QUARTERS
5¢	1 PENNY 1 NICKEL 2 QUARTERS
25¢	2 NICKELS
25¢	2 NICKELS

This puzzle requires the use of the nine assorted coins shown in the left-hand column. Each of these coins must be arranged to touch one or more of the other eight coins as designated in the right-hand column. Stacking coins or standing coins on edge is not allowed. All coins must remain flat on the tabletop. Only one "coinfusing" arrangement is possible.

Solution on page 82.

23 PERFECT PERFECT VISION

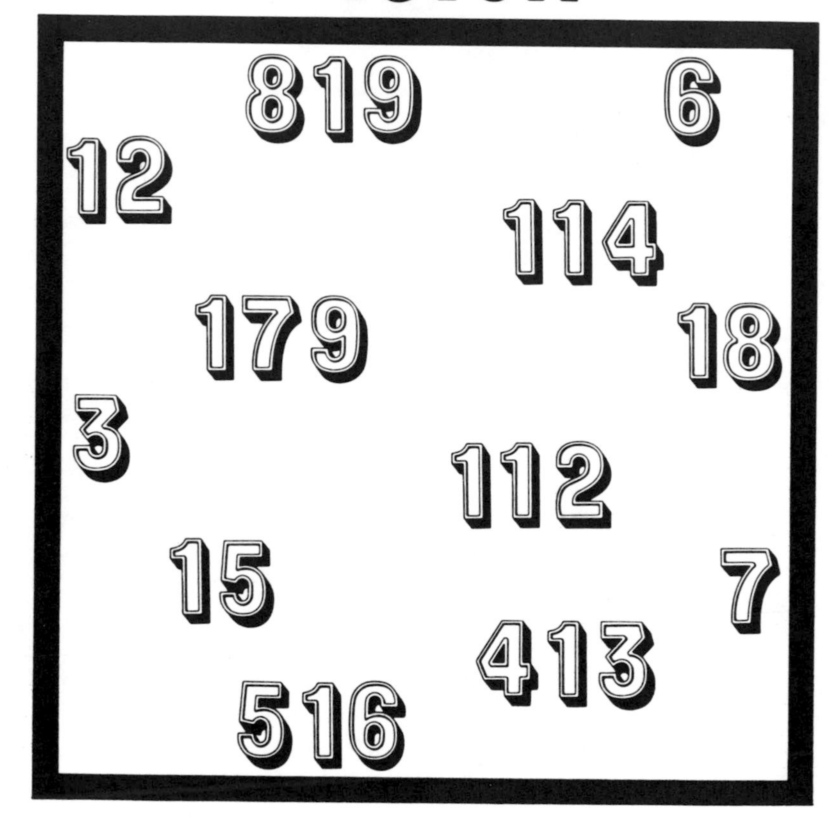

Using four straight lines, divide this square into nine pieces so that each piece totals the same number.

Solution on page 84.

24 Fair Is Wheel

Here is a puzzle that offers two ways to solve. Your challenge is to trace two paths that contain an "even" number of total rings beginning and ending at the double rings. Your first path must be the shortest possible route and your second path must be the longest possible route. No ring or connector may be used more than once per solution.

Solution on page 86.

25

PAINTIN' PLACE

A painter has five ten-gallon paint cans. From the following information, how many gallons of each color are left in each can? 1) There are eight gallons total of red and green paint. 2) There are ten gallons total of blue and yellow paint. 3) There are thirteen gallons total of green, yellow, and orange paint. 4) There are nine gallons total of red, blue, and orange paint.

Solution on page 88.

Clue to puzzle **16** *: Strangely, the area of the negative space created by the interior square will be greater than the area of the original shape.*

26

SEVENS THE HARD WAY

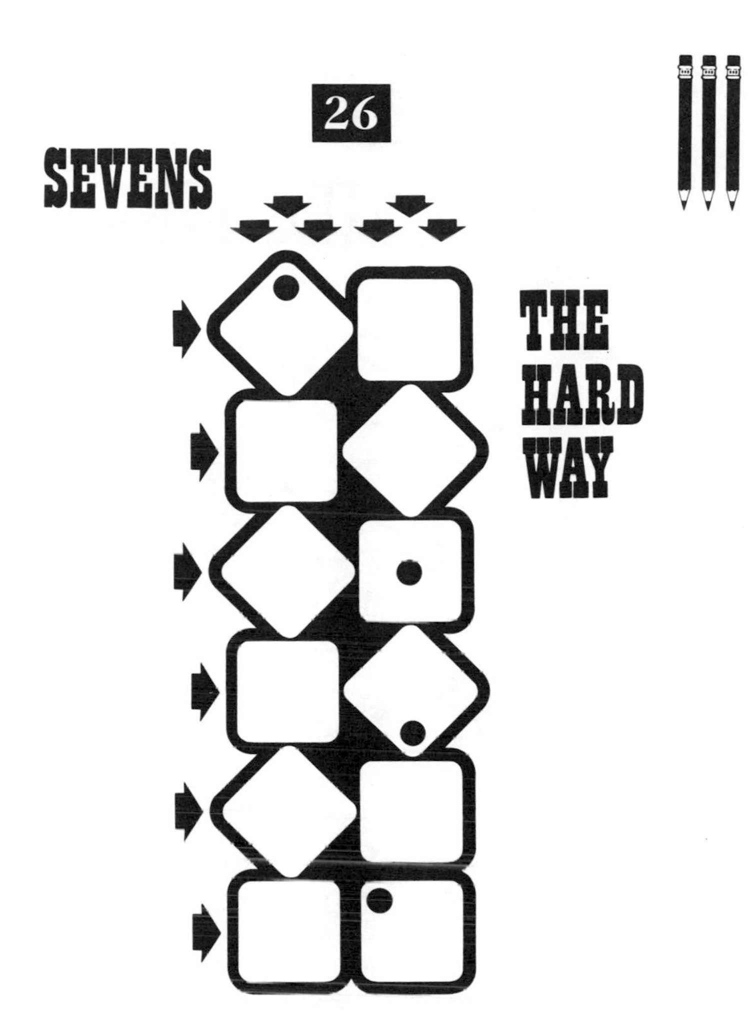

These six pairs of regulation dice have most of their pips missing. It is your task to position the missing pips to fulfill the following requirements: 1) Each of the six stacked pairs of dice must total seven. 2) Each of the six vertical pip columns pointed out by the six arrows at the top must also total exactly seven pips. 3) Both vertical towers of six dice must reveal all faces, one through six.

Solution on page 90.

27

PIE CENTENNIALS

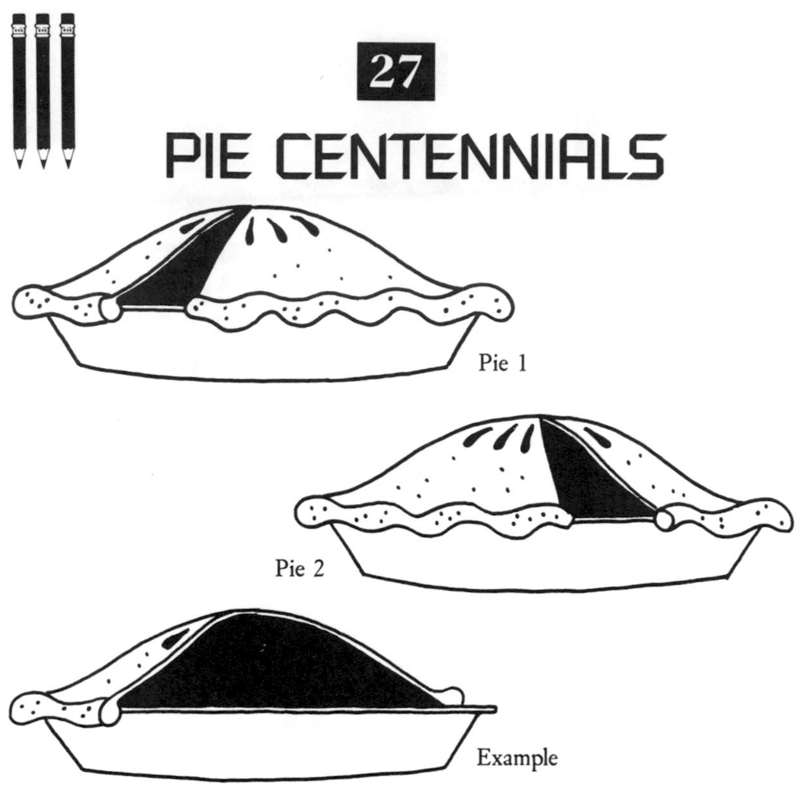

Each of the three pies above has been cut into two pieces with the smaller piece removed. Using the bottom pie as an example, the smaller piece measures $47\frac{3}{6}$ percent of the pie and the larger piece measures $52\frac{9}{18}$ percent. Together they total exactly 100 percent of the pie and utilize all nine numbers one through nine. Your task is to illustrate the piece sizes of the two remaining pies in the same manner. As in the example, all pieces must be made up of no less than four numbers and no more than five numbers. In the first pie, it is known that the smaller piece is less than five percent. In the second pie, it is known that the smaller piece is greater than eight percent.

Solution on page 92.

28

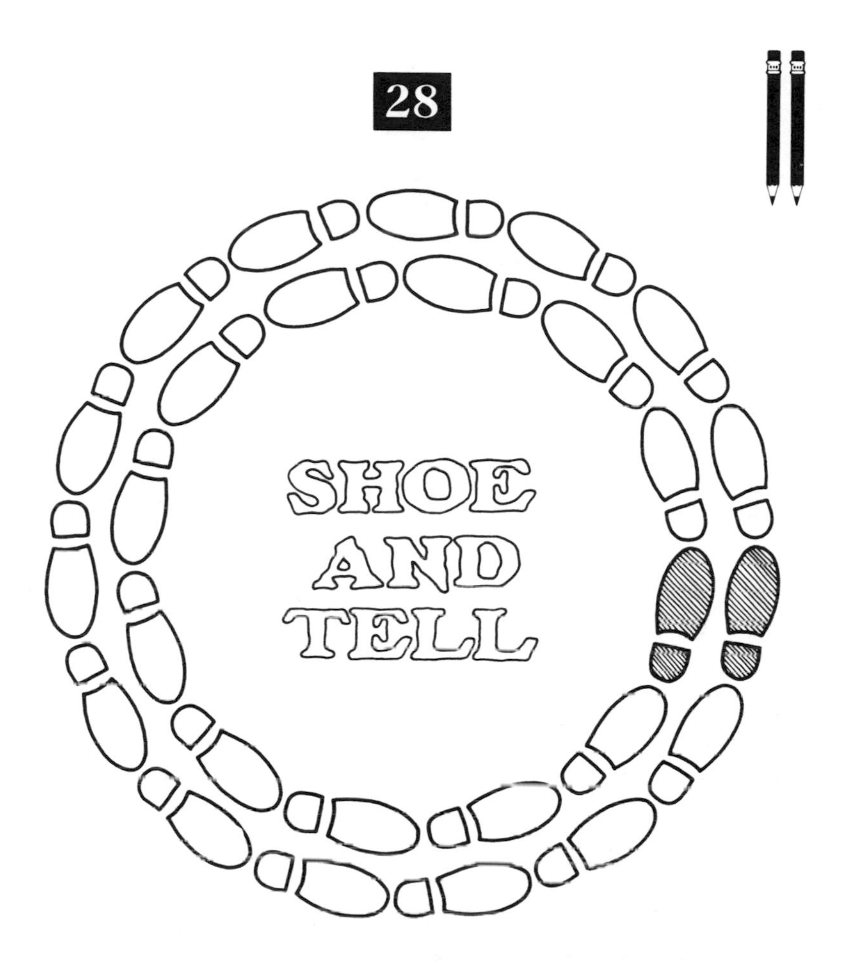

Here is a puzzle that will literally have you walking in circles. Start your circular stroll with both feet on the shaded footprints. Beginning with the right foot, walk in a normal manner, always stepping to the next marked print. Never skip a footprint. If we are told that each footstep measures 10 inches, how many steps are required before a stride of 52½ inches is reached?

Solution on page 94.

(Need a clue? Turn to page 35.)

29

STAR STUDIED

There are ten stars in this puzzle. You are to color eighty percent of the stars blue and seventy percent of the stars red. That means that fifty percent of the stars must be both blue and red. The stars must also be colored in such a way that two straight lines can be drawn to divide the stars into three groups: one in which the stars are all blue, one in which the stars are all red, and one in which the stars are both blue and red.

Solution on page 82.

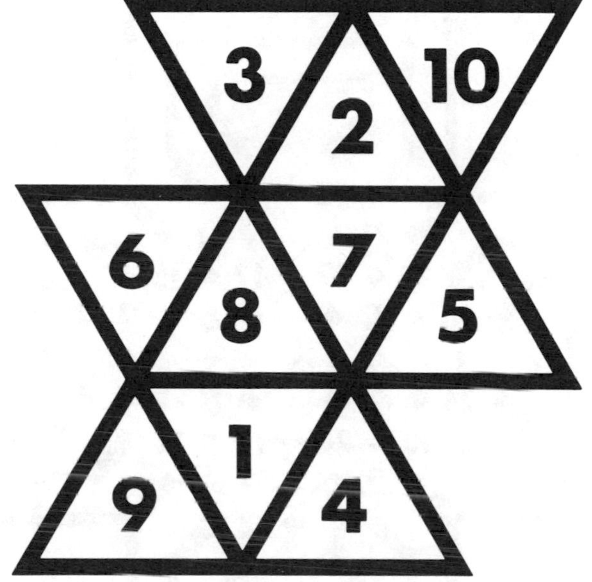

In this sample puzzle, the numbers one through ten have been so positioned that each group of four small triangles creates a larger triangle totalling twenty-two. There are four such overlapping triangles. It is your turn to find four additional solutions for this puzzle, each of which must have different totals.

Solution on page 84.

(Need a clue? Turn to page 44.)

31

WEAVE ONLY JUST BEGUN

The weave in this puzzle is not complete. When completed it will divide this puzzle into three separate paths. Eight key crossovers are already drawn in. Your task is to complete the remaining sixteen of the twenty-four overcrossings so that each strand will total exactly one hundred.

Solution on page 86.

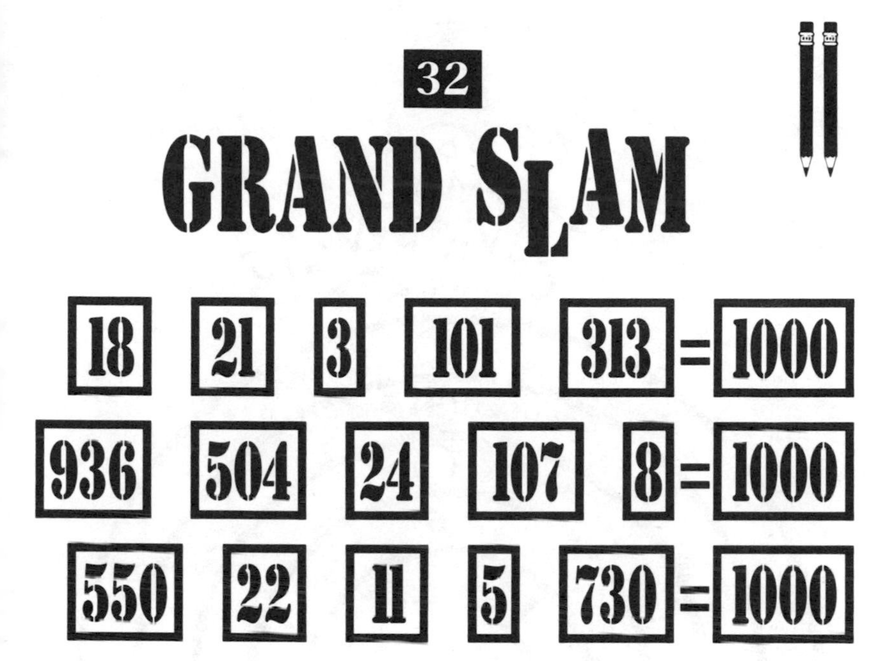

Before you are three separate challenges. It is your task to insert the four different mathematical operations (addition, subtraction, multiplication, and division) between the boxes in each equation in such a way that the final outcome is one thousand. You must perform each mathematical operation in the order in which it appears.

Solution on page 88.

Clue to puzzle **28** *: On the twenty-fourth step, the left foot has finished a cycle. The right foot is three prints short of a cycle. Therefore, a stride of 30 inches has been reached. This means each step taken by the left foot increases the stride by 2½ inches, arrived at by dividing 30 inches by twelve steps.*

A sample solution is given here, but others remain. Starting at the arrow, as you travel inward to the center of the spiral, drop off the numbers one through eight consecutively at the circles of your choice. Upon reaching the end of the spiral, each of the four rows of circles must total nine. Find the five remaining solutions.

Solution on page 90.

34

MAGIC WORD SQUARES

Each letter in this puzzle represents a different number from zero to nine. It is your challenge to switch these letters back to numbers in such a way that each horizontal, vertical, and diagonal row of three words totals the same number. Your total for this puzzle is 1515.

Solution on page 92.

(Need a clue? Turn to page 47.)

35

DISC COUNTS

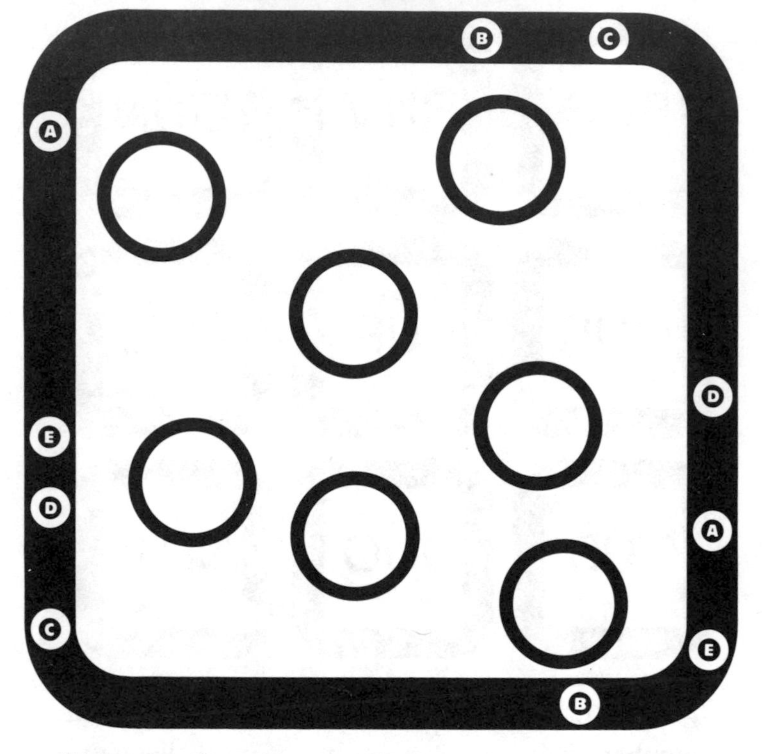

To begin, draw five straight lines that connect identical letters located on the perimeter of the puzzle (A to A, B to B, etc.). Now, position the numbers one through seven in the seven discs so that each of the five lines totals the same number.

Solution on page 94.

(Need a clue? Turn to page 50.)

36

TRIPLE PLAY

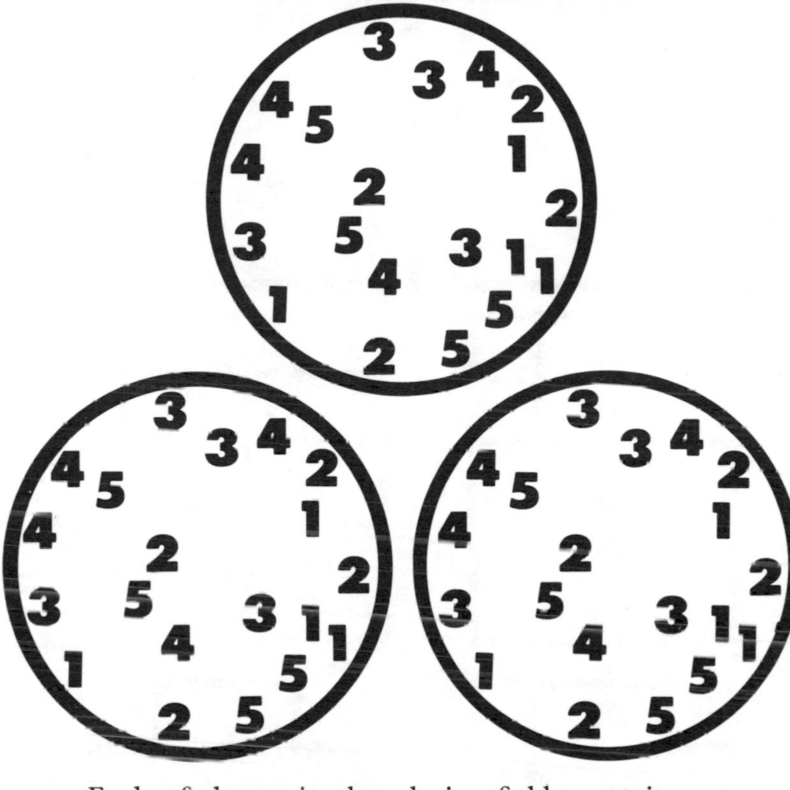

Each of these circular playing fields contains an identical arrangement of numbers. Your challenge is to solve each circle in a different manner. 1) Using two straight lines divide the first circle into three pieces which each total twenty. 2) Using two straight lines divide the next circle into four pieces which each total fifteen. 3) Using three straight lines divide the last circle into five pieces which each total twelve.

Solution on page 82.

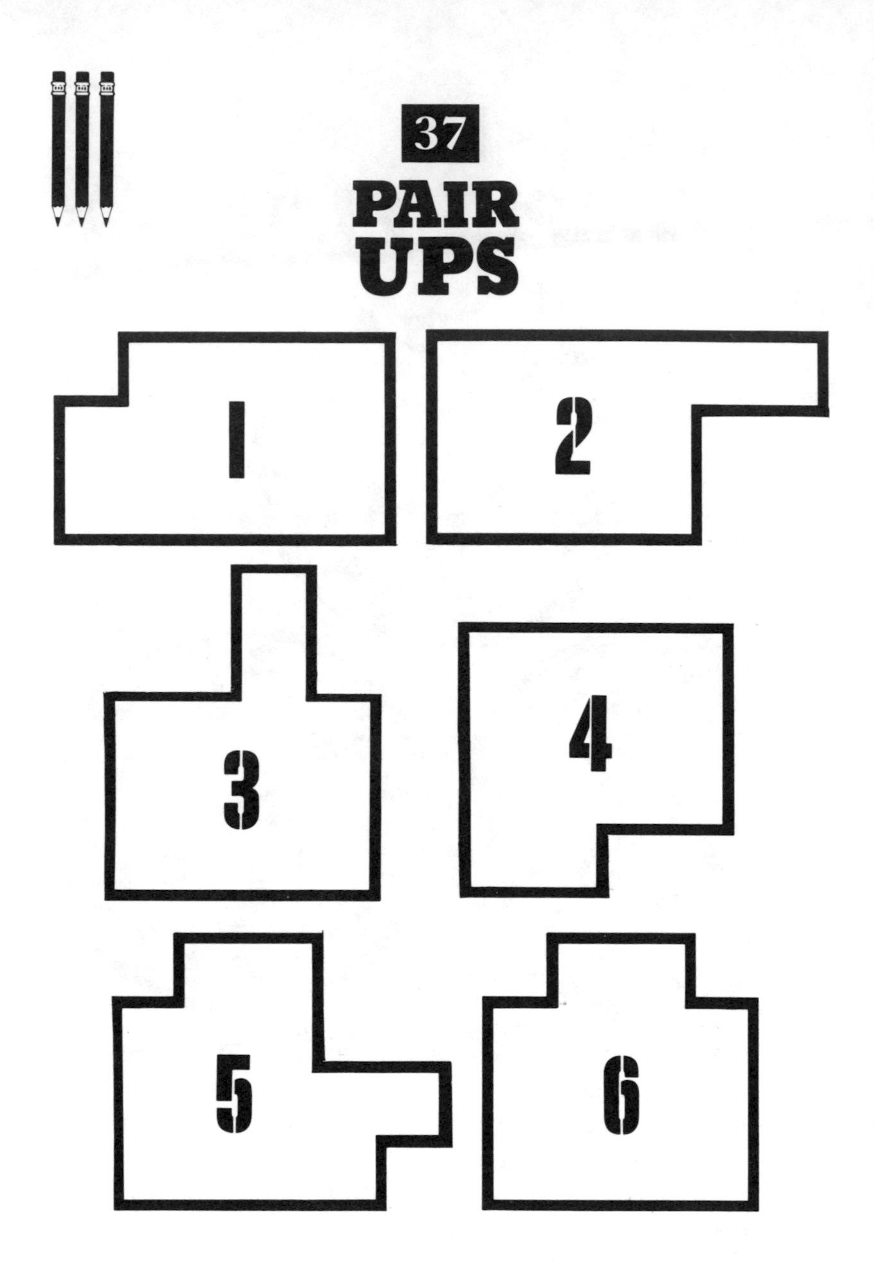

Pair up the six shapes in this puzzle to form three pieces of equal size and shape.

Solution on page 84.

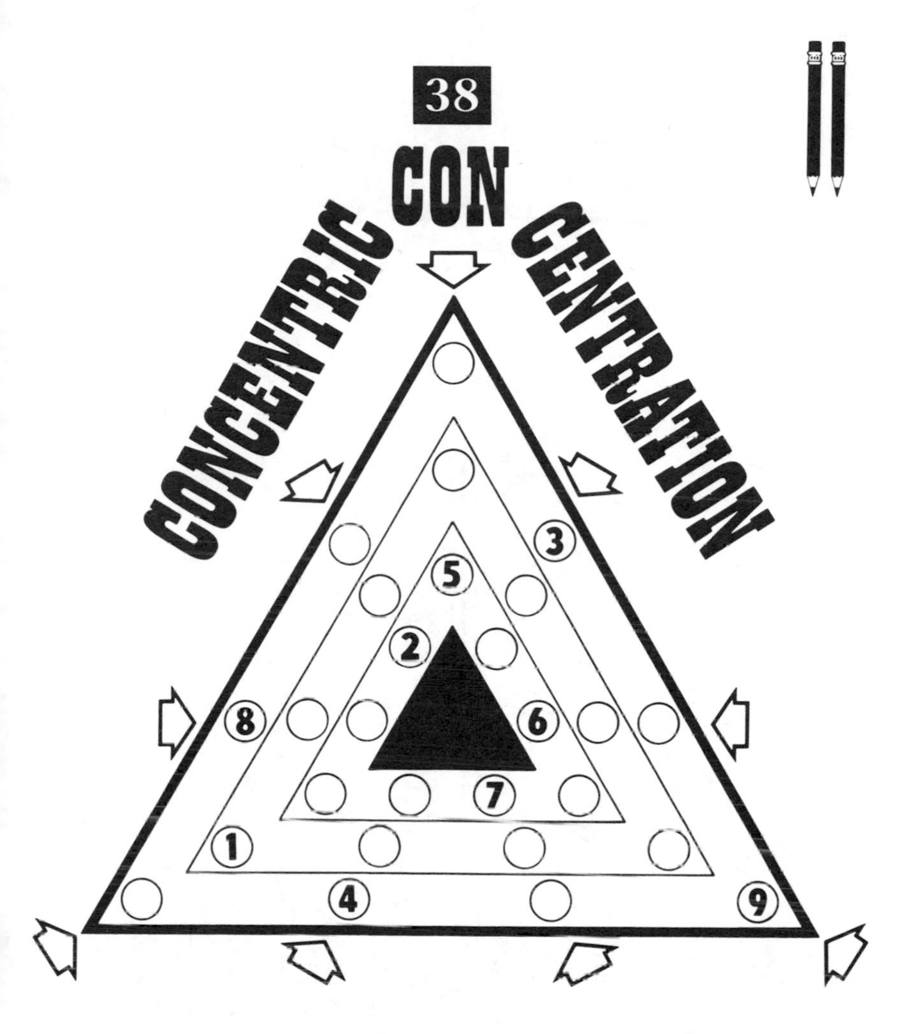

Each of the concentric triangles in this puzzle sports nine circles. As a head start, nine numbers have already been positioned in these circles. Position the missing numbers in the following manner: 1) The numbers one through nine must appear in each triangle. 2) Each side of each triangle (four circles) must total twenty. 3) Each row of three numbers illustrated with arrows must total fifteen.

Solution on page 86.

39

OPTICAL DIVERSION

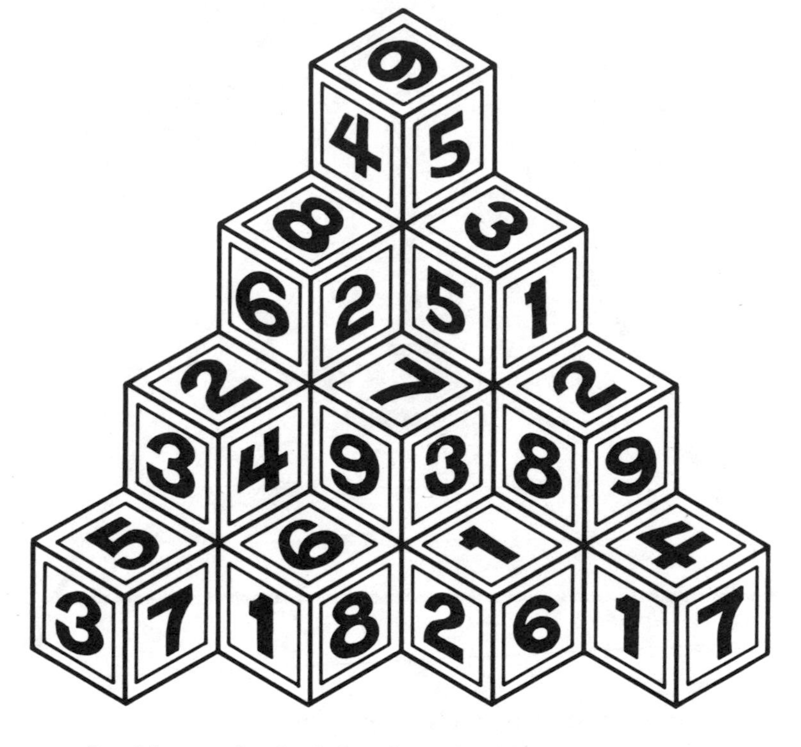

In this puzzle, find the three building blocks which contain the numbers one through nine. To reveal your answer simply black out all unnecessary parts of the pyramid.

Solution on page 88.

(Need a clue? Turn to page 55.)

40

CRISSCROSS CRISIS

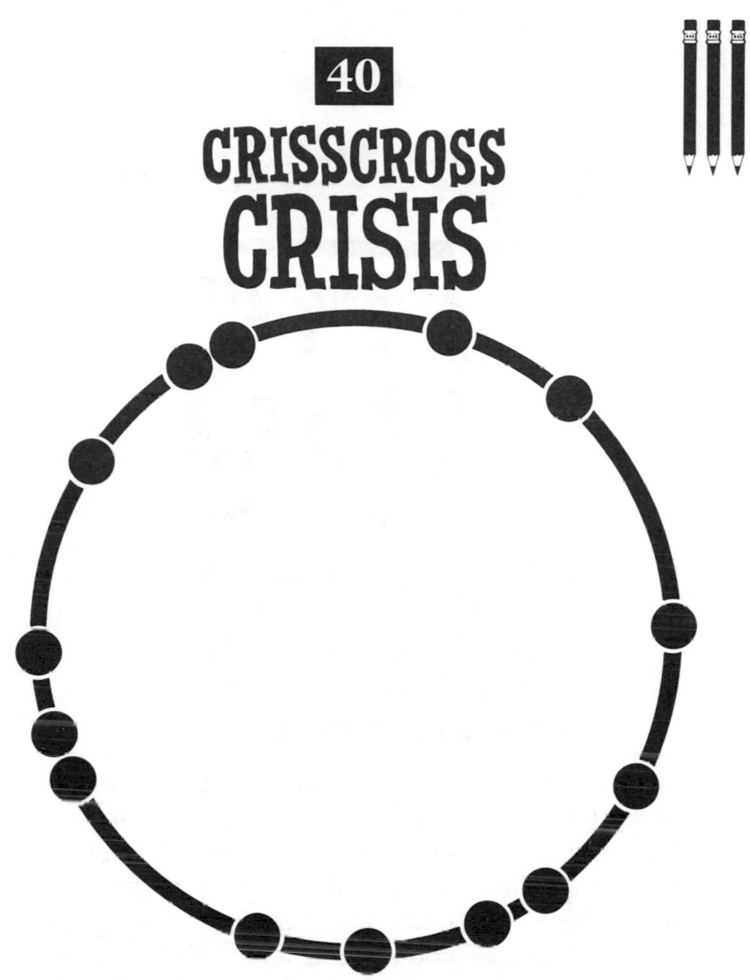

Create the largest number of intersections possible (twenty-one) in the interior of this circle by drawing seven straight lines that utilize all fourteen spots on the perimeter. Each line must connect two spots.

Here is a second challenge: How many spots are required before ten times this number of intersections are possible? This figure can be calculated without drawing the circle.

Solution on page 90.

41

AMPLE TURNOVERS

Carefully examine the two three-digit numbers in this puzzle. When added together they total 795. Now, turn the page upside down and you'll see they now total 1590, that's exactly twice the original total of 795. Can you find another pair of three-digit numbers which produce this same phenomenon using two new totals? Just as in the example you may not repeat numbers in a single figure and your numbers may never begin with a zero.

Solution on page 92.

Clue to puzzle **30** *: The four totals are eighteen, twenty, twenty-four, and twenty-six.*

42

Trimming the Tree

This Christmas tree has been decorated with a stringer of thirty-one lights. The color of four bulbs has already been determined. It is known that one of the light sockets is broken. You must determine which socket is broken so that the remaining colored lights can be arranged in the following manner: 1) Use an equal number of red, blue, yellow, orange, and green lights. 2) Position these lights on the stringer in a repetitive order (example: R-B-Y-O-G, R-B-Y-O-G, etc.). 3) The five lights along each side of the tree must include all five colors.

Solution on page 94.

43

SIDESTEP

Here is a strategic paper-and-pencil game for two players: to win the game, simply create a box canyon which allows you to make the last move. Play begins at any square and continues with players alternating moves to any open adjacent horizontal, vertical, or diagonal square. In the example game above, you are the odd player. It is your move and challenge to set up a box canyon which allows you to win on move eleven.

Solution on page 83.

44

FIGUREHEADS

Figure "A" has discovered that he can create a five-digit number which is exactly five times greater than a five-digit number that figure "B" can create. However, figure "B" has discovered that his number can be twice the size of the number figure "A" can create by simply eliminating A's zero. What are the numbers that "A" and "B" are thinking of?

Solution on page 85.

Clue to puzzle **34**: *BUG = 602.*

45

ROMAN STONE GAME

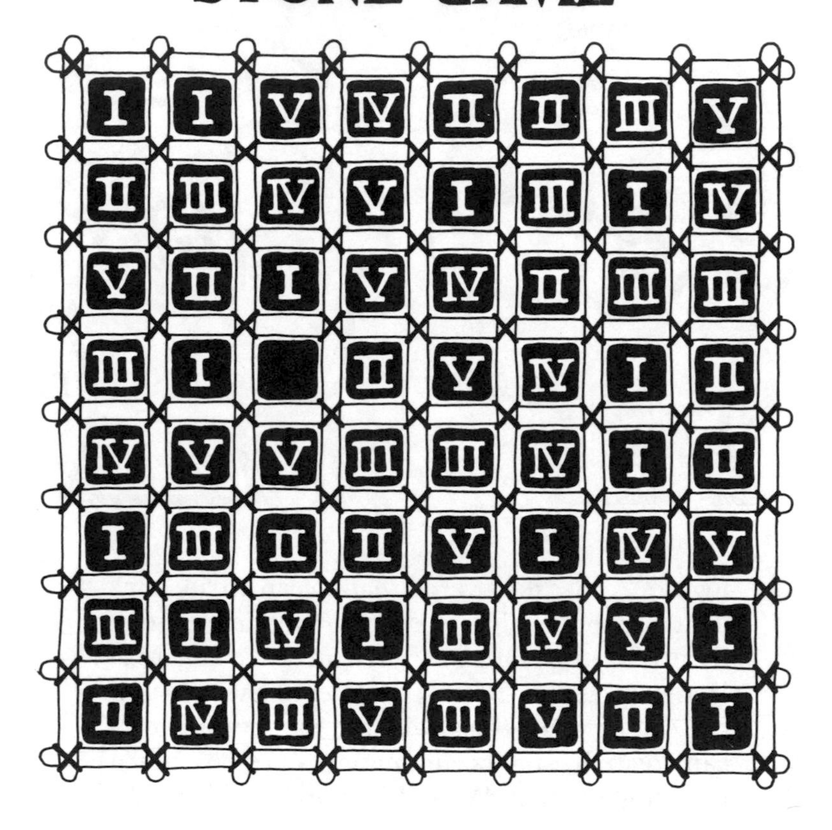

Blacken the surface of twenty-four stones in this puzzle in such a way that the eight horizontal and eight vertical rows will each contain five stones of different numeric value. One stone has already been blackened for you.

Solution on page 87.

46

TRAIL 'N' ERROR

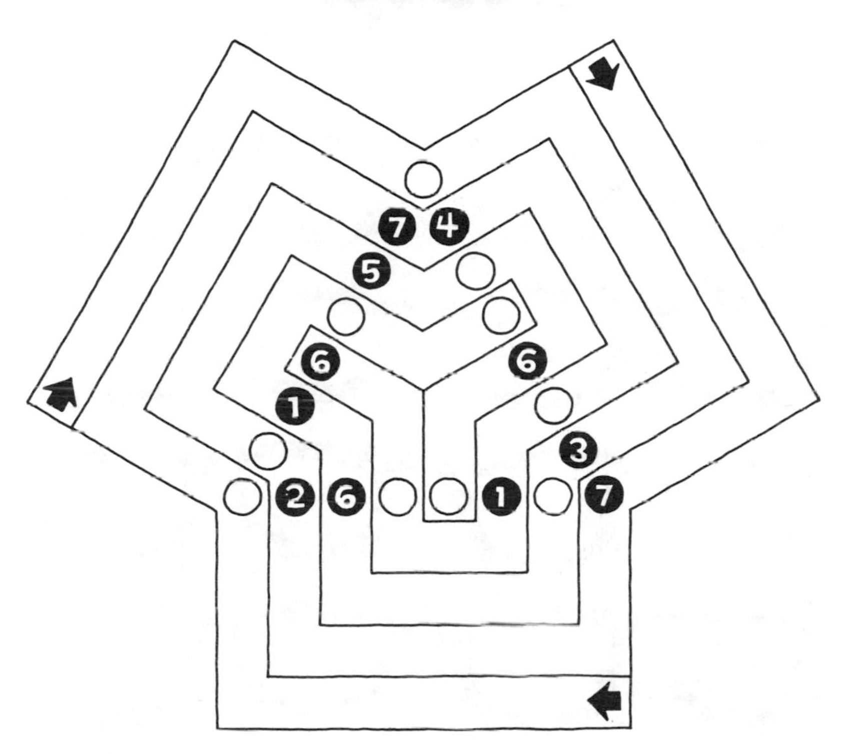

Each spiralling trail in this puzzle must contain the numbers one through seven. Eleven numbers have already been positioned. Arrange the missing numbers in such a way that each side of the equilateral triangle will total thirty-two.

Solution on page 89.

47

Up to Specs

Position the numbers one through fifteen on the fifteen unbroken lenses in this puzzle. Do this in such a way that each pair of spectacles totals fifteen and each horizontal, vertical, and diagonal row of four lenses totals thirty.

Solution on page 91.

48 Graduate School

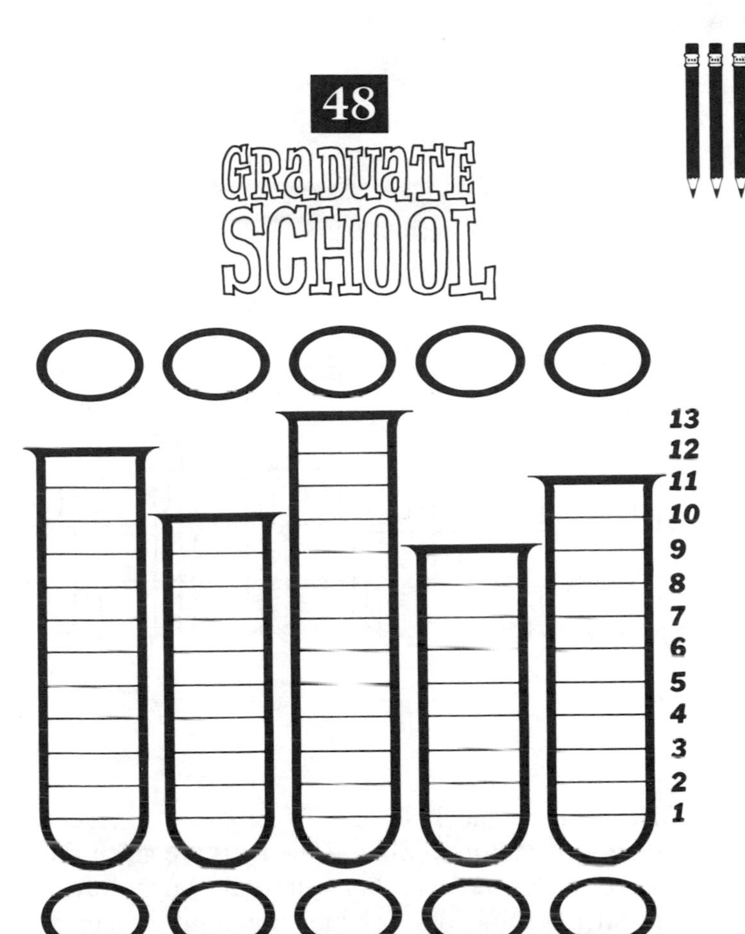

These five test tubes are graduated in one-ounce increments as marked on the right. The ovals below the tubes are for charting the ounces of water in each. The ovals above are for charting the unfilled ounces of each tube. Your challenge is to fill the five test tubes with a total of twenty-nine ounces of water to fulfill the following criteria: 1) From left to right each test tube must contain more water. 2) The ovals above and below must divulge the numbers one through ten.

Solution on page 93.

49

BAR BELLS

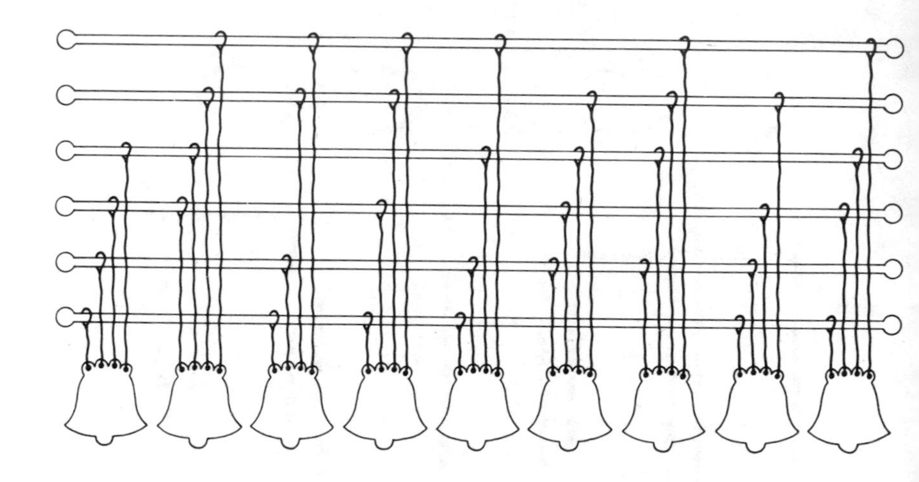

Give each of the bells in this puzzle a different number from one to nine. If the bells are given the correct numbers, each bar will total exactly thirty. The strings show to which bars the number in each bell must be added.

Solution on page 83.

(Need a clue? Turn to page 63.)

Clue to puzzle **35** *: Triskaidekaphobes, beware!*

50

NEVER SAY DIE

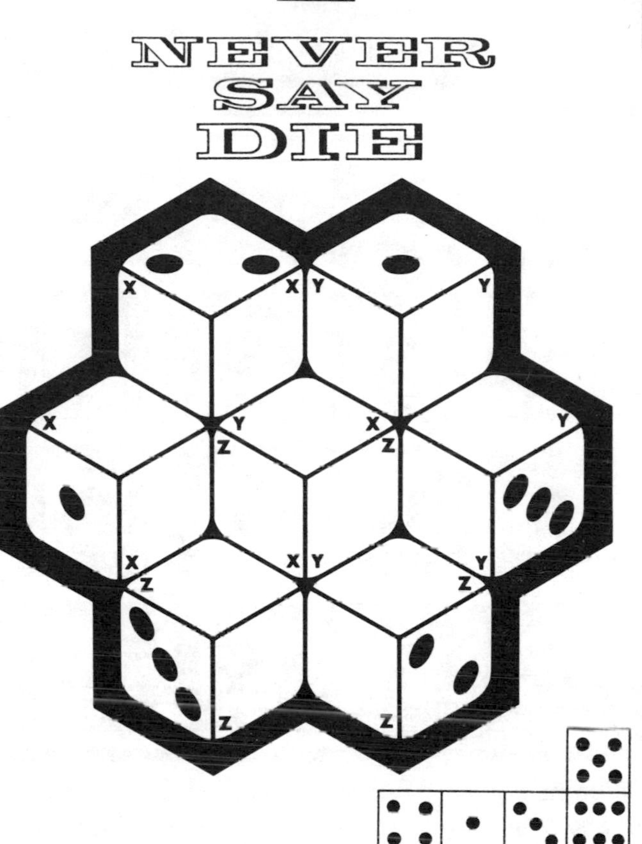

Position the missing pips on the seven standard dice of this puzzle. Do this in such a way that each of the three six-pointed stars which are identified by the letters "X," "Y," and "Z" contain all six pip patterns. The illustrated foldout of a standard die must be used in solving the puzzle.

Solution on page 85.

51

BEHIND CLOSED DOORS

Opening one of the numbered doors shown cancels out two numbers, the number on the door itself and the door covered by the opened door. (For example, opening the first door cancels both the 2 on the door itself and the 6 on the door below, which it covers.) In compensation, behind each door is a number, revealed when the door is opened, that is half the value of the number on the door. Can you open four doors so that all the horizontal and vertical rows add up to the same total?

Solution on page 87.

(Need a clue? Turn to page 64.)

52

The Olympi-Add

Place the numbers one through nine in the nine divided areas that make up these Olympic Rings so that the sum of any pair of overlapping rings will total twenty-two. There are two possible solutions.

Solution on page 89.

Clue to puzzle **39** *: Instead of a bird's-eye view (from above), the cubes you seek defy gravity to give you a worm's-eye view (from below).*

53
abstract treason

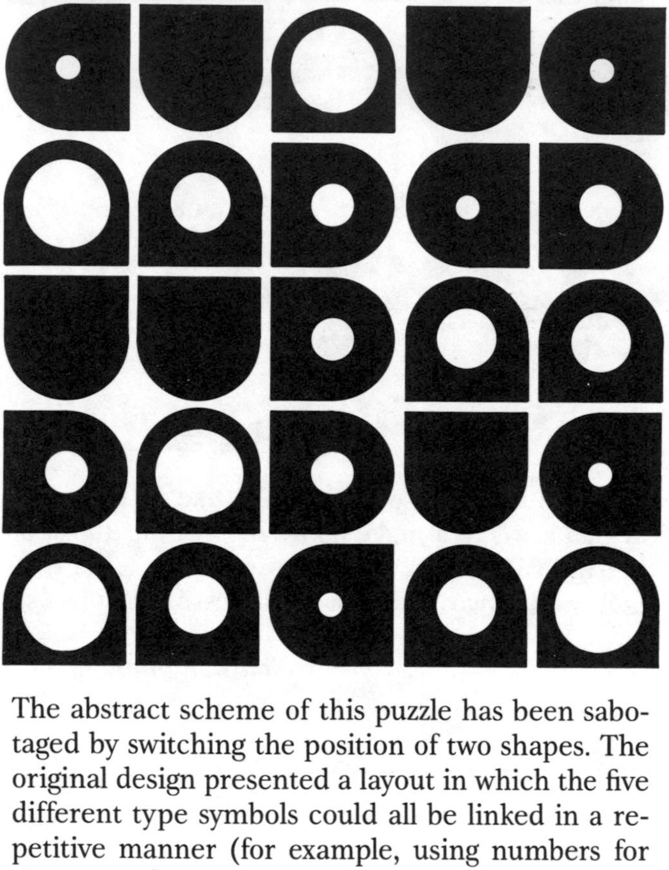

The abstract scheme of this puzzle has been sabotaged by switching the position of two shapes. The original design presented a layout in which the five different type symbols could all be linked in a repetitive manner (for example, using numbers for shapes, 1-2-3-4-5, 1-2-3-4-5, etc.) with a single continuous line travelling horizontally, vertically, and diagonally from one adjacent shape to another. Your challenge is to restore the original design scheme to the puzzle and trace the path that links all of these shapes.

Solution on page 91.

Splice the six clips of film in this puzzle to produce one fifteen-frame short feature. The final result must be a strip which presents a mathematically logical number sequence.

Solution on page 93.

55

MAGIC WORD SQUARES

ORB	TIP	PIT
PEE	ORA	TWO
TOW	POD	OAR

Each letter in this puzzle represents a different number from zero to nine. It is your challenge to switch these letters back to numbers in such a way that each horizontal, vertical, and diagonal row of three words totals the same number. Your total for this puzzle is 1167.

Solution on page 83.

(Need a clue? Turn to page 68.)

56 GLOBE TROTTER

You have one year to tour the globe (365 days). During your tour, twenty-seven stops must be made (symbolized by the twenty-seven squares encircling the globe). To complete this tour, you must create a schedule by selecting two numbers which signify both the number of stops you will make and the total number of days that will be spent at each stop. (Example: If one of your numbers is ten, this means that you must make ten stops and spend ten days at each stop for a total of one hundred days.) Your task is to find two numbers which create a combined total of twenty-seven stops and use a total of 365 days.

Solution on page 87.

57

NAVAHO RUG

Divide this rug into nine pieces of equal size and shape. The design element within each shape must be identical.

Solution on page 85.

58

Tacks Reform

A thrifty yet geometrically esthetic carpenter has discovered a way to use only six tacks to secure all nine pieces of wood in such a way that no single piece of wood can slide. He positioned the tacks to create four equilateral triangles. How did he do it?

Solution on page 89.

59

WHIPPER SNAPPER

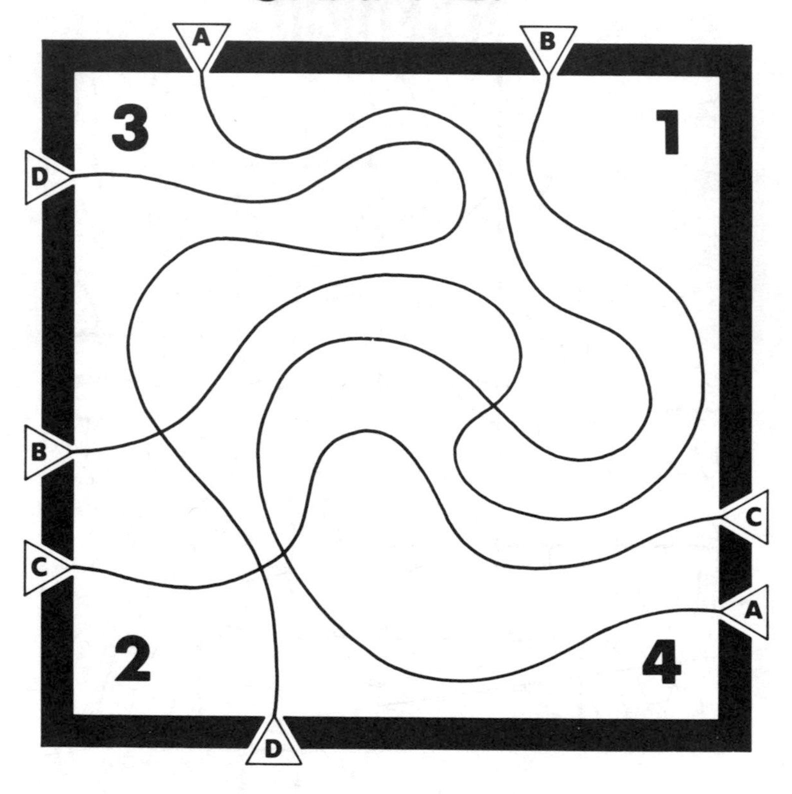

Four whips (AA, BB, CC, and DD) divide this puzzle into nine irregular shapes. There are six shapes which border each whip. The numbers one through nine must be placed in these shapes to assure that all shapes which border each individual whip will total exactly thirty. The numbers one through four have been positioned for you.

Solution on page 91.

60

TALLY OF THE DOLLS

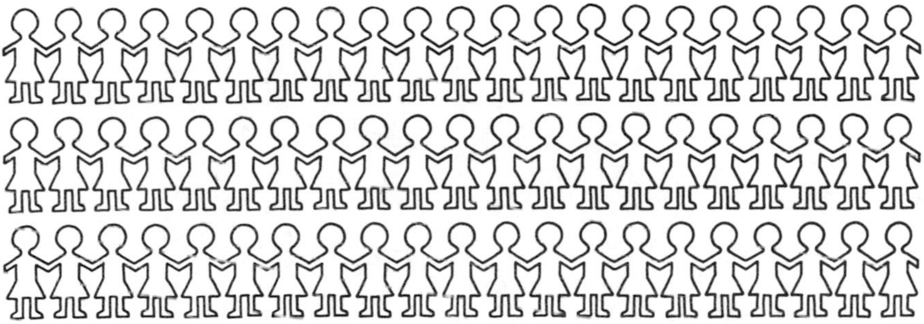

For each row of twenty-one paper dolls, select a number and make two cuts dividing the row into three groups so that, by using the number twice in the blanks, you complete (and illustrate) the following mathematical statement:

The tally of the dolls in the left group is ______ times larger than the tally of the dolls in the middle group; the tally of the dolls in the middle group is also ______ times larger than the tally of the dolls in the right group.

Solution on page 93.

Clue to puzzle **49** *: Ring up a 3 on the first bell.*

SIX SHOOTER

There are twenty-six gun sights in this puzzle. With the aid of a compass or piece of string, draw six arcs of equal radius which pass through the cross hairs of each gun sight.

Solution on page 83.

Clue to puzzle **51** *: Each horizontal and vertical row must total twenty.*

There are two ways to solve this problem of topology. Using the fewest number of colors, color all fifteen hexagons to fulfill the following requirements: 1) No adjacent hexagon may be of the same color. 2) All hexagons of identical color must be an equal distance apart from one another.

Solution on page 85.

(Need a clue? Turn to page 72.)

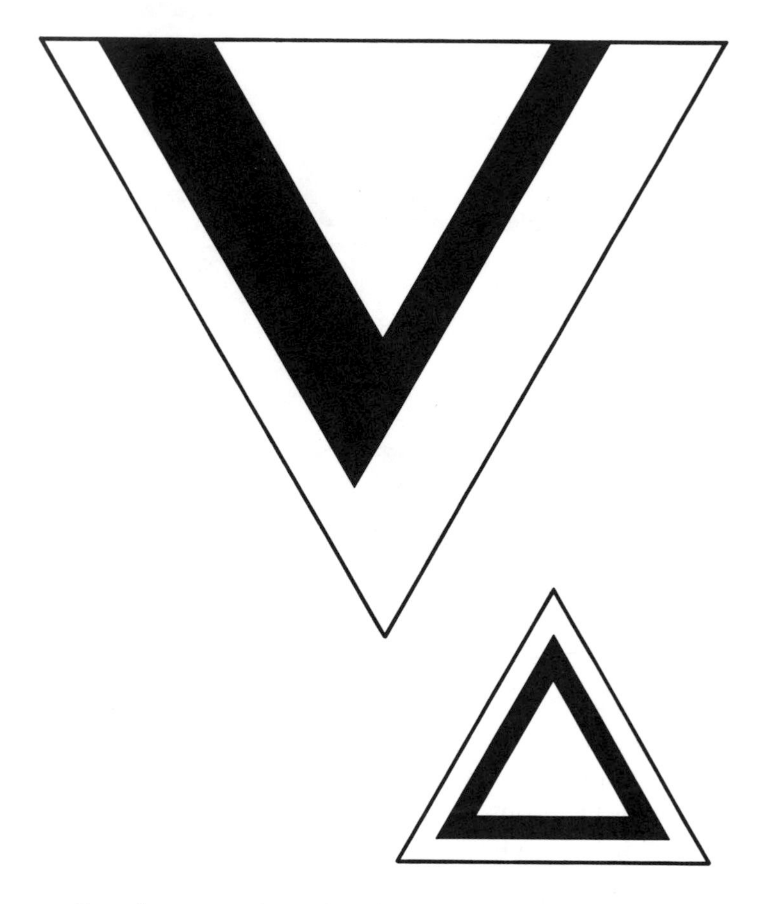

Cut the top triangular figure into three V-shaped pieces which can be reassembled to form the bottom figure, which is shown in reduced scale.

Solution on page 87.

64

COUNT CADENCE

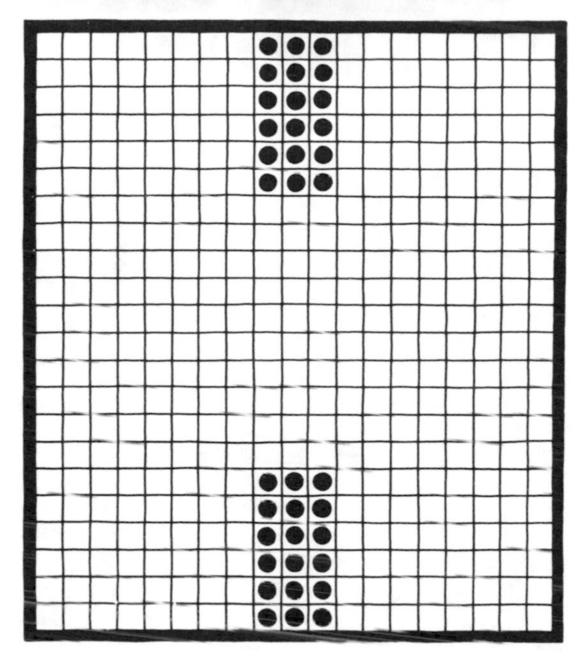

Eight different commands are necessary to maneuver the eighteen-man drill team in this puzzle from the position below to the position above. It is known that the first command is, "Nine squares forward march." The remaining seven commands are listed in a scrambled manner below:

Fourteen squares left flank march.
Ten squares to the rear march.
Five squares left flank march.
Six squares right flank march.
Thirteen squares to the rear march.
Four squares right flank march.
Seven squares left flank march.

Solution on page 89.

65

DEVIOUS DIVIDERS

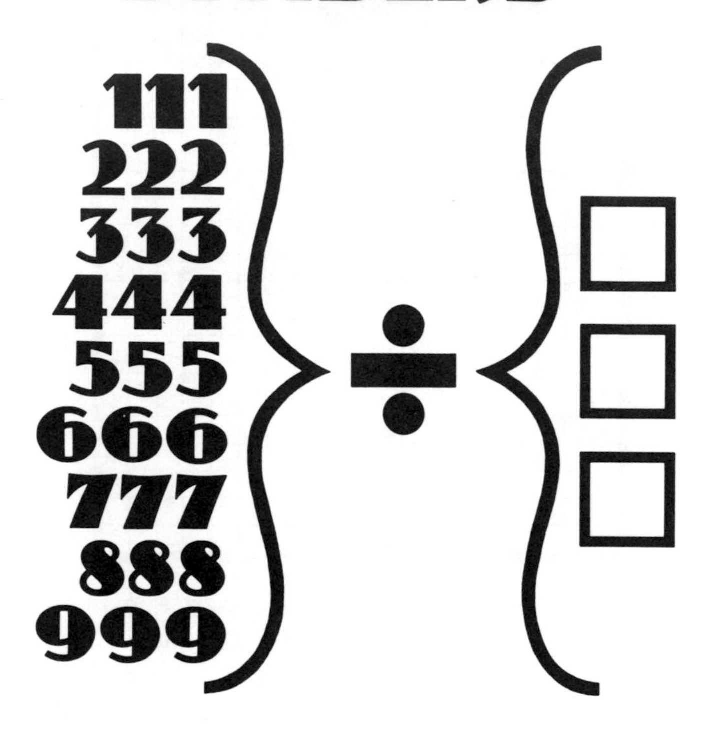

Besides the number one, can you find three whole numbers which will divide evenly into each of the nine three-digit numbers in this puzzle?

Solution on page 91.

Clue to puzzle **55** *: ORB = 380.*

66

COMMON CENTS

Coin	Touches
1¢	2 DIMES
1¢	2 QUARTERS
5¢	2 DIMES 2 QUARTERS
10¢	1 PENNY 2 QUARTERS 1 NICKEL
10¢	1 PENNY 2 QUARTERS 1 NICKEL
25¢	1 PENNY 1 DIME
25¢	1 PENNY 1 DIME
25¢	1 NICKEL 1 DIME
25¢	1 NICKEL 1 DIME

This puzzle requires the use of the nine assorted coins shown in the left-hand column. Each of these coins must be arranged to touch one or more of the other eight coins as designated in the right-hand column. Stacking coins or standing coins on edge is not allowed. All coins must remain flat on the tabletop. Only one "coinfusing" arrangement is possible.

Solution on page 93.

67

TAKE FIVE

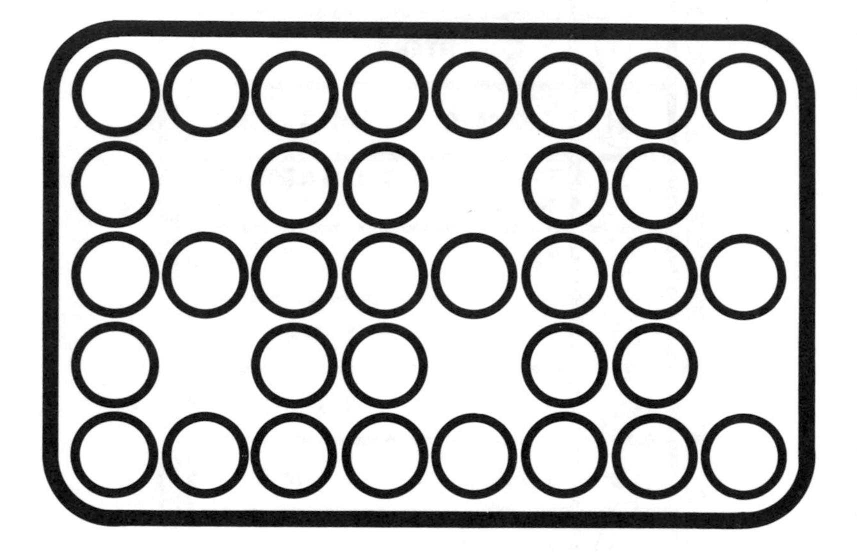

There are thirty-four circles in this puzzle. Your challenge is to remove five circles to leave five.

Solution on page 83.

(Need a clue? Turn to page 74.)

68

STRIP PROVOKER

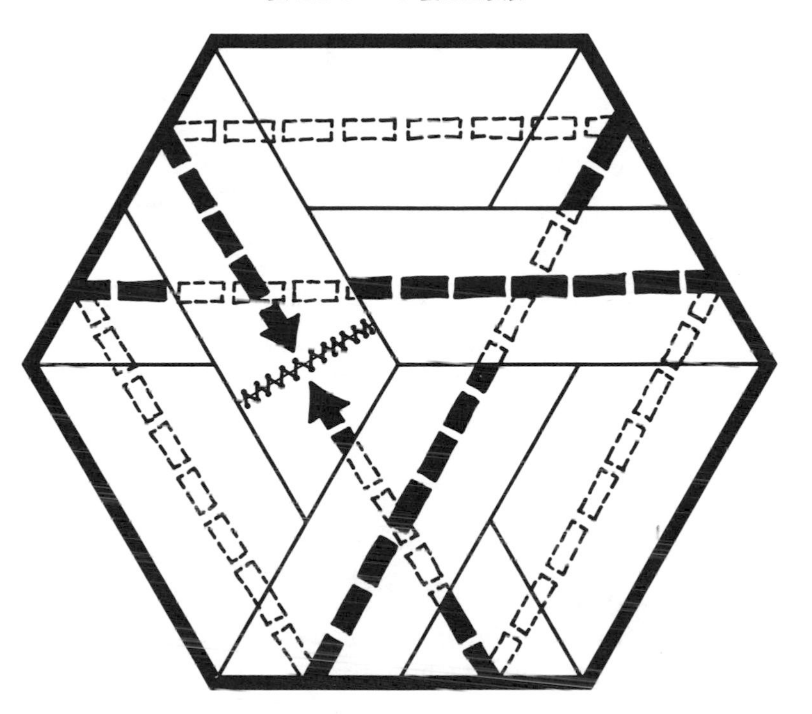

Illustrated in this puzzle is a single strip of paper which has been folded and stitched together at both ends. Find the easiest way to determine the approximate total length of the strip. It is known that the width of this strip is 7/8 of an inch.

Solution on page 87.

69

GRAND SLAM

568	71	4	103	236	=	1000
24	618	15	9	70	=	1000
203	65	64	168	9	=	1000

Before you are three separate challenges. It is your task to insert the four different mathematical operations (addition, subtraction, multiplication, and division) between the boxes in each equation in such a way that the final outcome is one thousand. You must perform each mathematical operation in the order in which it appears.

Solution on page 85.

Clue to puzzle **62***: Each color must form an equilateral triangle.*

70

GEOMETRACTS

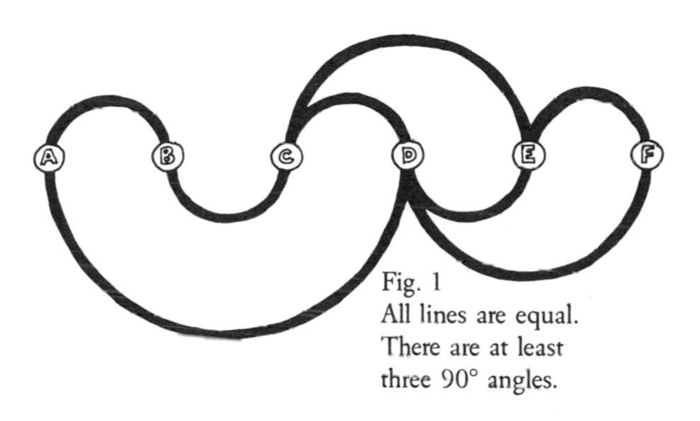

Fig. 1
All lines are equal.
There are at least
three 90° angles.

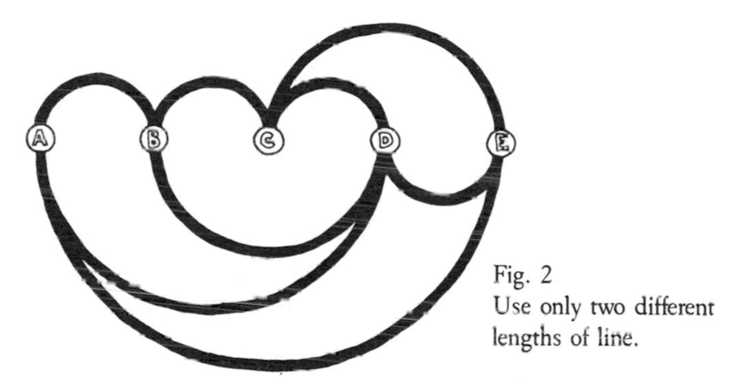

Fig. 2
Use only two different
lengths of line.

Here are two distorted geometric figures. Both have been stretched in such a way that the original figure is unrecognizable at first glance. Your task is to straighten all the lines in each figure to reveal its original identity. The circled letters designate the intersection of two or more lines. Vital clues are given for each figure.

Solution on page 89.

71

TOOTH PACE

This puzzle will help you brush up on your mechanical aptitude. Rotate the large gear on the right in the direction of the arrow until the number three is in the position now occupied by the number four. You must then calculate which three teeth will mesh together at each of the five gear junctures.

Solution on page 91.

> *Clue to puzzle* **67** *: Darkening various circles may literally help you see things more clearly.*

72

STOP AND LOOK LESSON

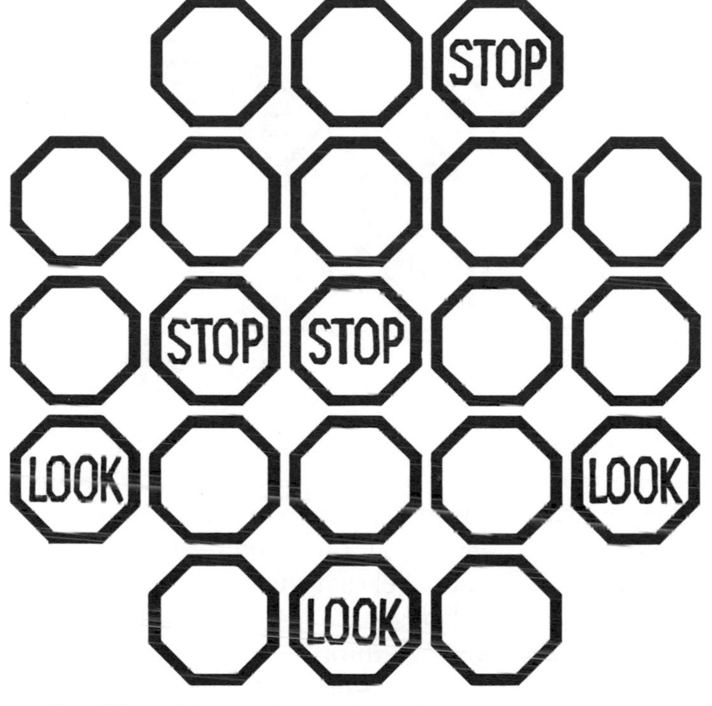

Partition this configuration of twenty-one octagons into three individual pieces which each contain seven octagons. Now, separate the pieces and rearrange them in such a way that the original configuration of octagons is reconstructed, with these exceptions: 1) The stop signs must be adjacent and located in a straight line. 2) The look signs must be adjacent and located in a straight line.

Solution on page 93.

73

THE MATHEMATICIAN'S FIGURE-EIGHT CAKEWALK

Illustrated here is a thirty-one-square mathematician's cakewalk in the shape of a figure eight. All figure-eight cakewalks are numbered in the following manner: Starting at the center square zero, move one step and label that square "1," move two steps and label that square "2," and so on, proceeding consecutively, until all squares are filled. Continue moving in the same direction and always travel straight through the intersection. Although previously numbered squares may be stepped on more than once, they may never be landed upon again. In the world of mathematics, less than twenty such cakewalks can be constructed using the first one million numbers in our number system. See if you can create the next-smaller and the next-larger figure-eight cakewalks.

Solution on page 83.

74

CONSECUTIVE DECISION

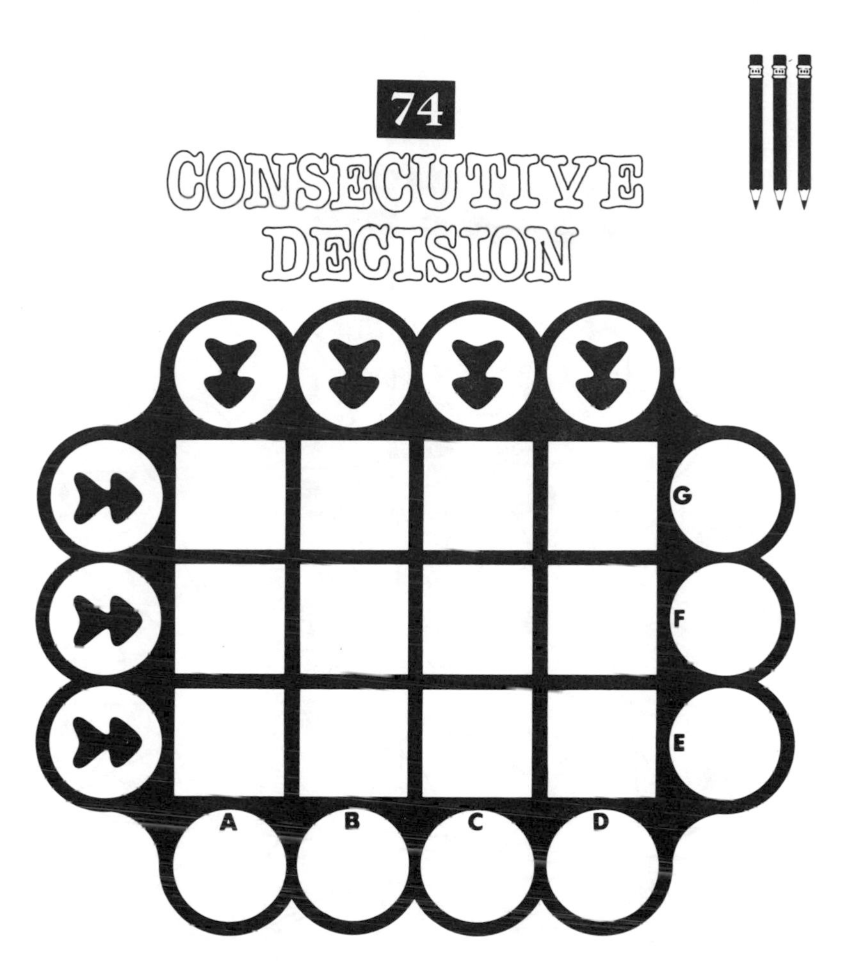

Beginning at the letter "A," position a series of seven consecutive numbers from lowest to highest in the seven circles. Next, select another series of twelve consecutive numbers and place them in the twelve squares so that the sum of each horizontal and vertical row will total the number in the respective circle.

Solution on page 85.

(Need a clue? Turn to page 79.)

75

SUNSTROKE

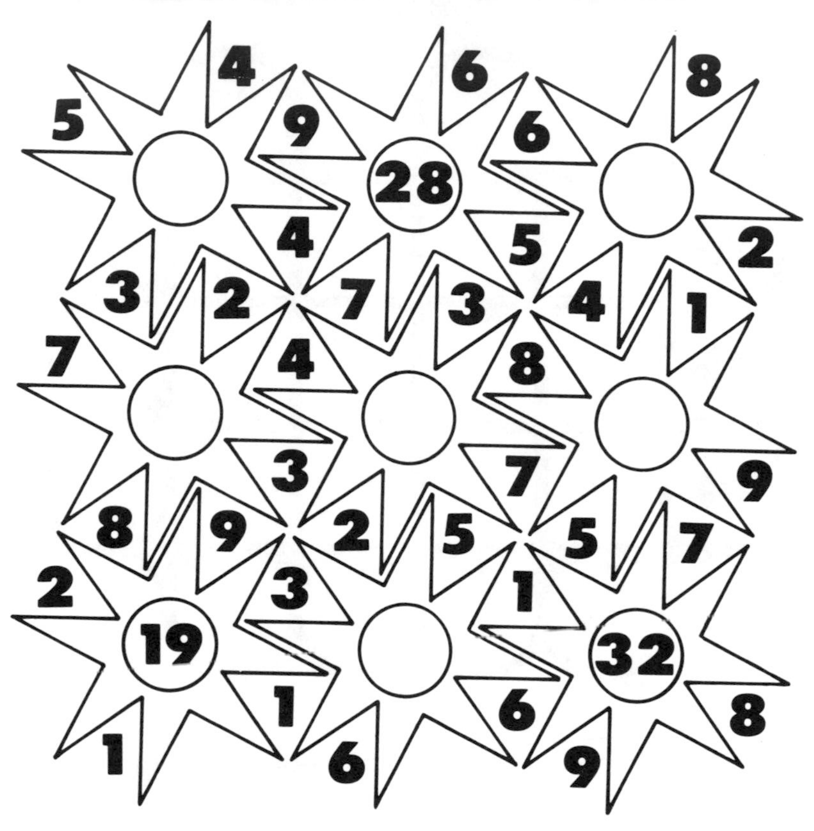

Fill in the six numbers which are missing from the empty sunspots. To get you started in the right "direction," three sunspot totals have been positioned for you. All thirty-six of the satellite numbers which hover around the sunspots determine the nine sunspot totals.

Solution on page 87.

76

RING AROUND THE ROSES

There are fifteen roses in this puzzle. It is your challenge to draw four circles which will enclose an odd number of roses. Also, each circle must enclose a different number of roses. No circle may touch or intersect another circle.

Solution on page 89.

Clue to puzzle **74** *: Nine is A good place to start.*

77

THE CARPENTER & THE GLASS CUTTER

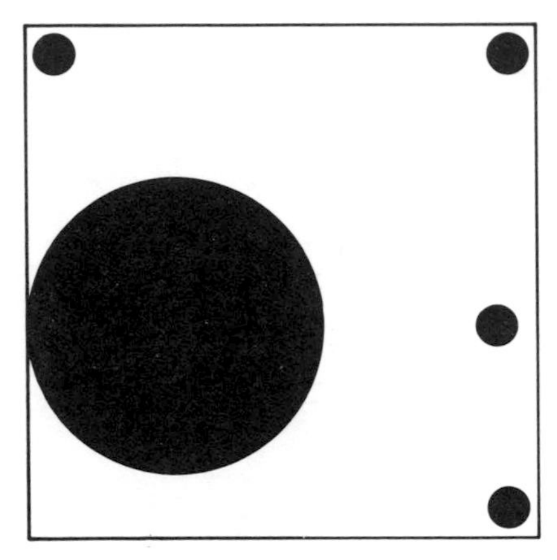

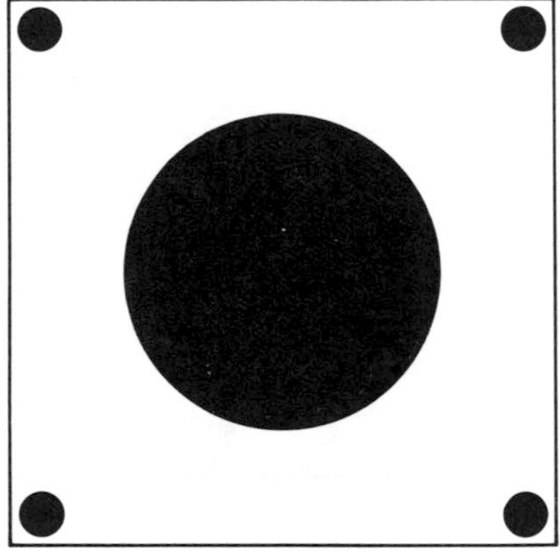

In this puzzle you must determine the best solution for two craftsmen; one is working with wood and the other is working with glass. Using the fewest straight cuts per craftsman, cut the top square figure into the fewest number of pieces which can be rearranged to form the bottom square figure. The white area is transparent for the glass cutter.

Solution on page 91.

78

STAR OF DIVIDE

Using only four straight lines, divide this six-pointed star into four sections, each containing one of the mathematical symbols. Create these divisions in such a way that, when the mathematical symbol is performed with the number three and the number of triangles in that section, all of the answers will be the same. In other words, add three to the number of triangles within the addition segment, subtract three from the number of triangles in the subtraction segment, multiply by three the number of triangles in the multiplication segment, and divide three into the number of triangles in the division segment to achieve the same number.

Solution on page 93.

Solutions

1 Knotholes 49, 34, and 17.

22

8

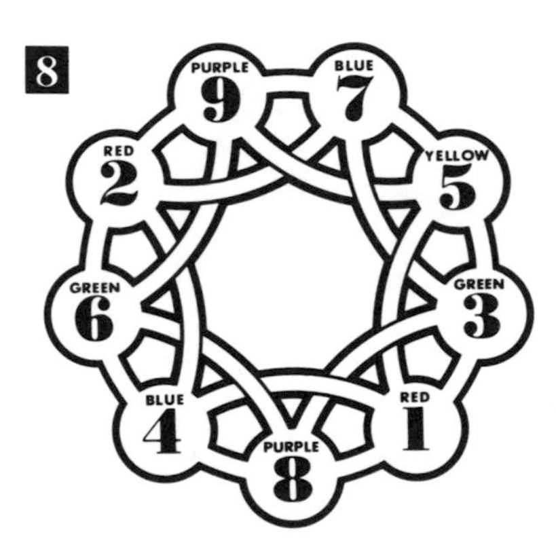

29

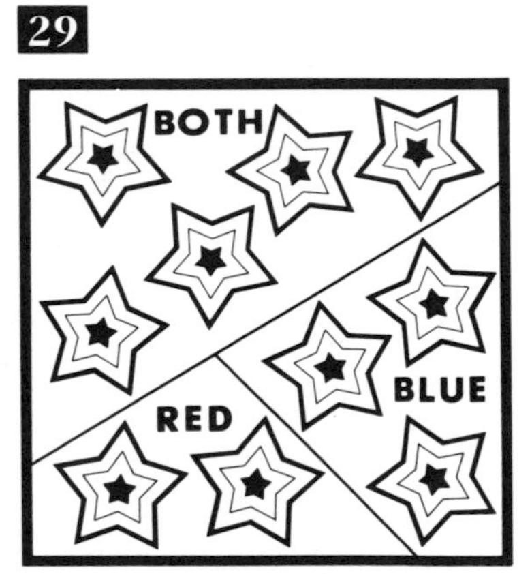

15

	1	
2	3	
4	5	6
9	7	8
11	10	12
13	14	15
39	40	41

36

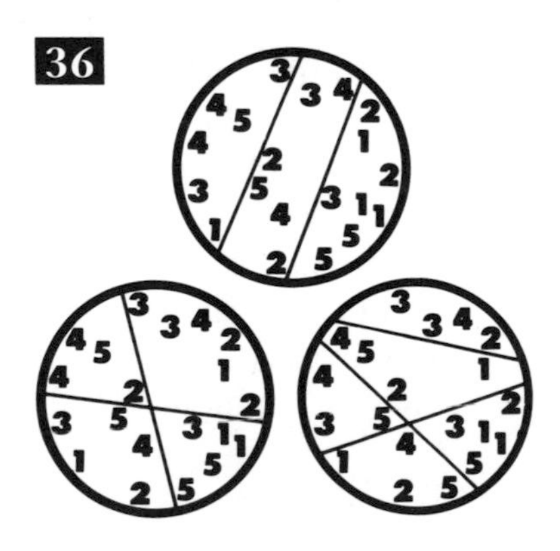

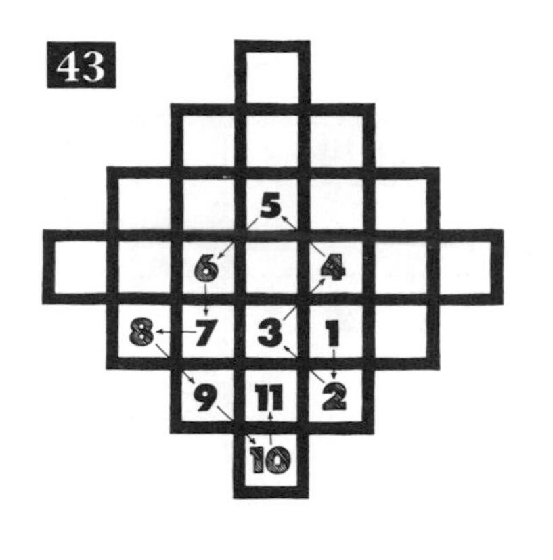

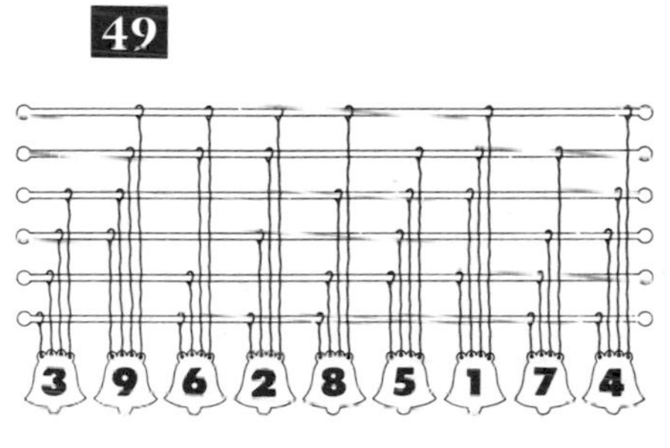

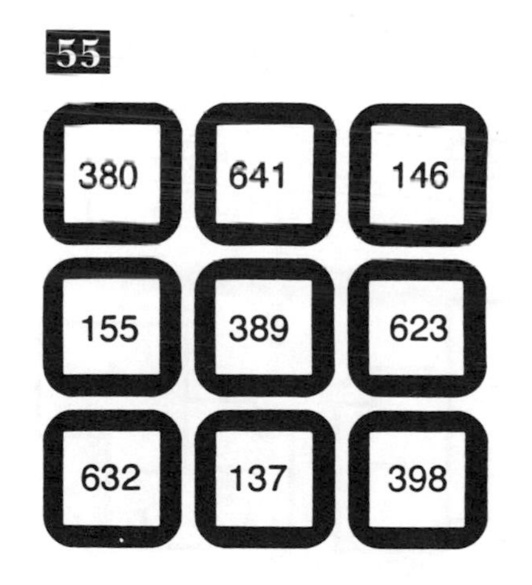

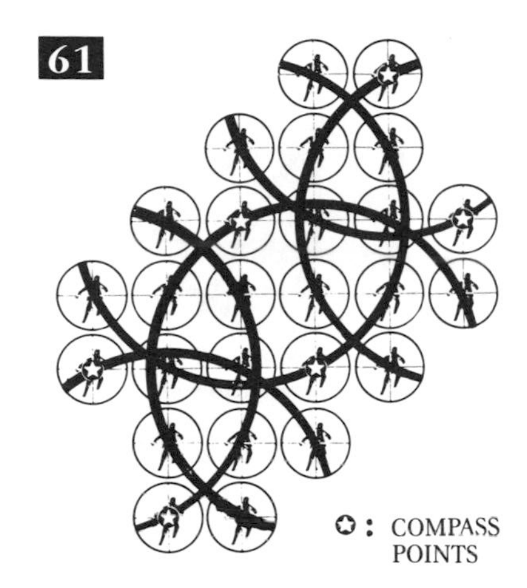

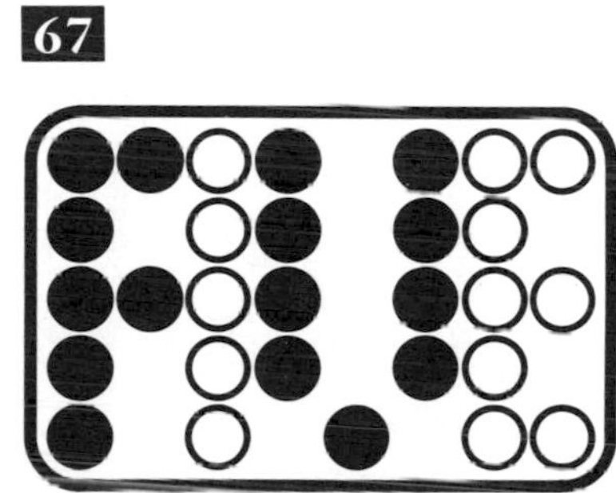

73 The cakewalks are fifteen and sixty-three squares, respectively. There are an infinite number of larger cakewalks. Each increasing cakewalk is double in size plus one square.

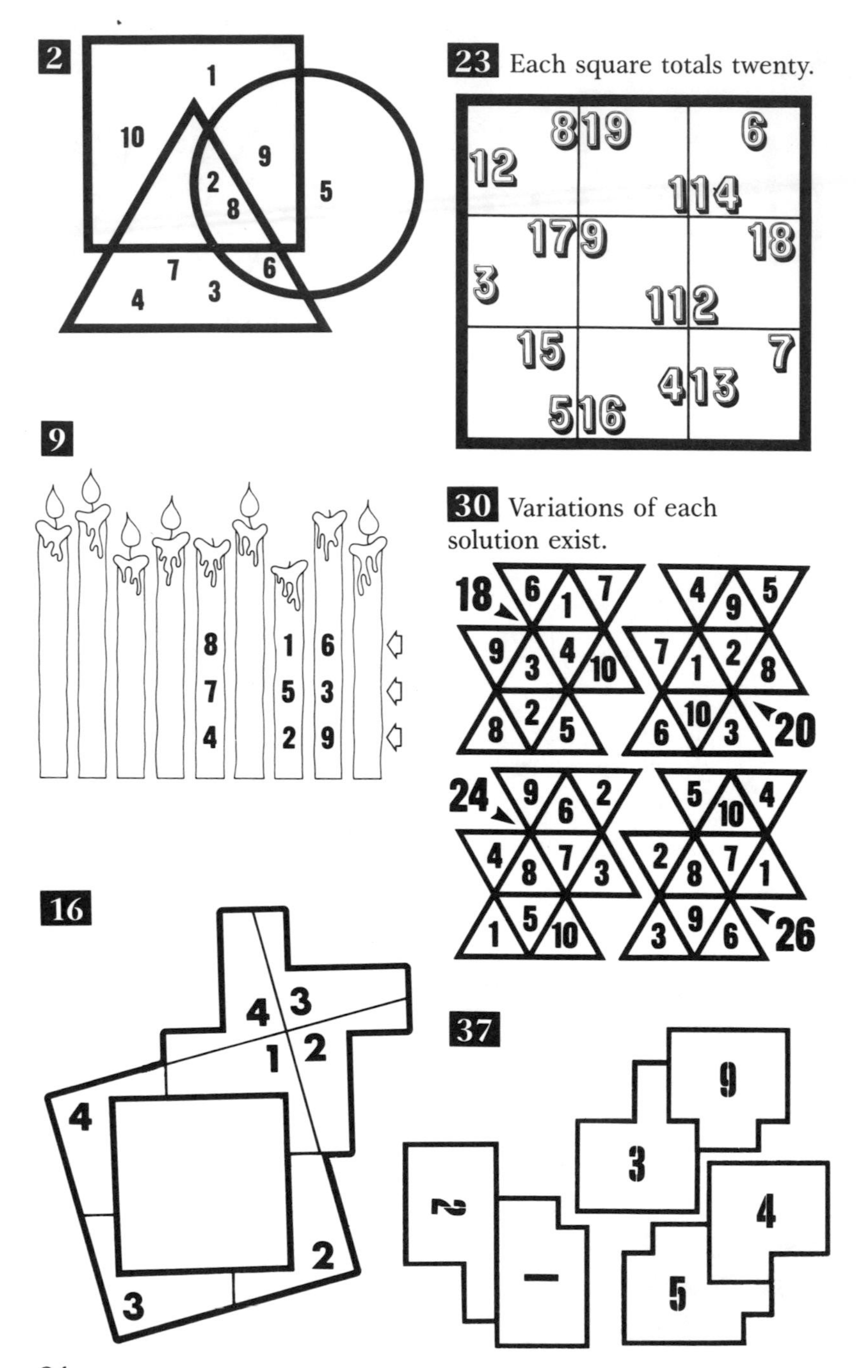
2
1
10
9
2
8
5
7
6
4
3
23 Each square totals twenty.
8 19
6
12
1 14
17 9
18
3
1 12
15
7
4 13
5 16
9
8
7
4
1
5
2
6
3
9
30 Variations of each solution exist.
18
6
1
7
9
3
4
10
8
2
5
4
9
5
7
1
2
8
6
10
3
20
24
9
6
2
4
8
7
3
1
5
10
5
10
4
2
8
7
1
3
9
6
26
16
4
3
1
2
4
2
3
37
9
3
2
1
4
5

44 The number 72,930 is five times 14,586, and 14,586 is twice 7,293.

50

57

62 Each number represents a different color. Mirror images also correct.

69

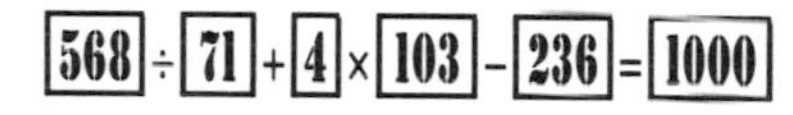

74

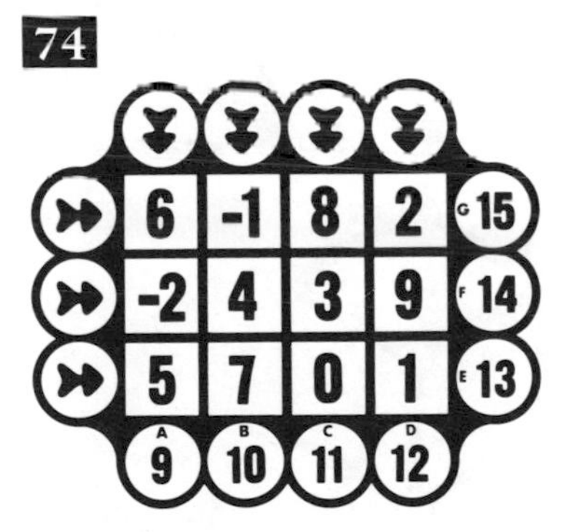

3

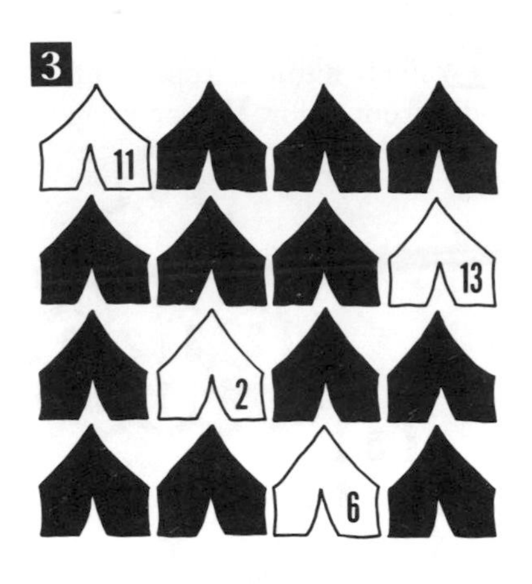

10

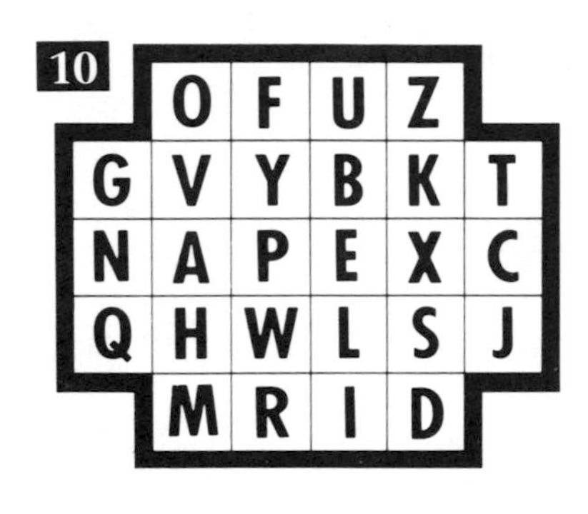

17

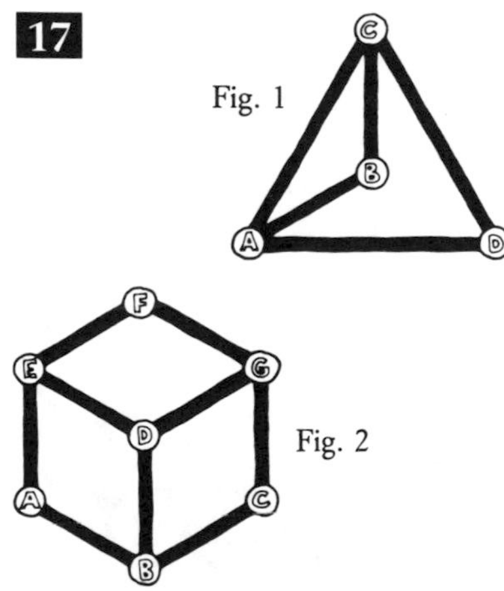

24 Below is one of several maximum routes. The shortest route is 1, 10, 11, 12, 13, 9, 27, 28, 31, 32.

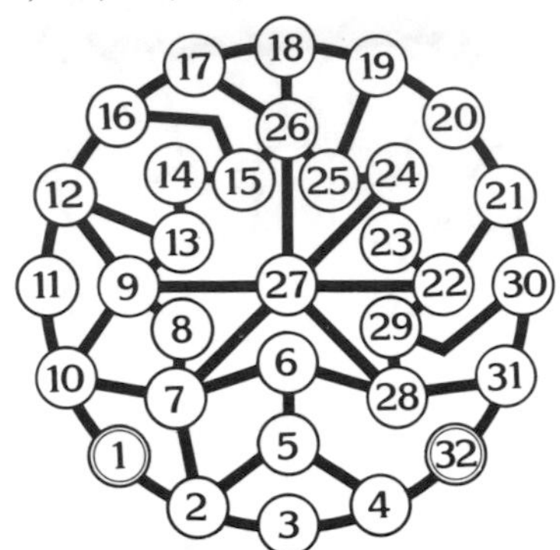

31

38

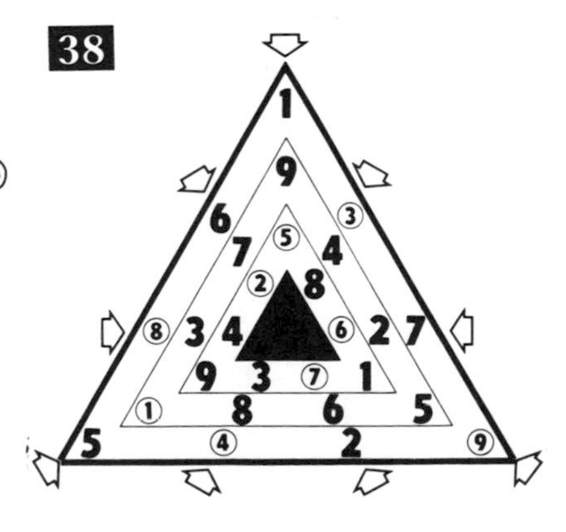

45

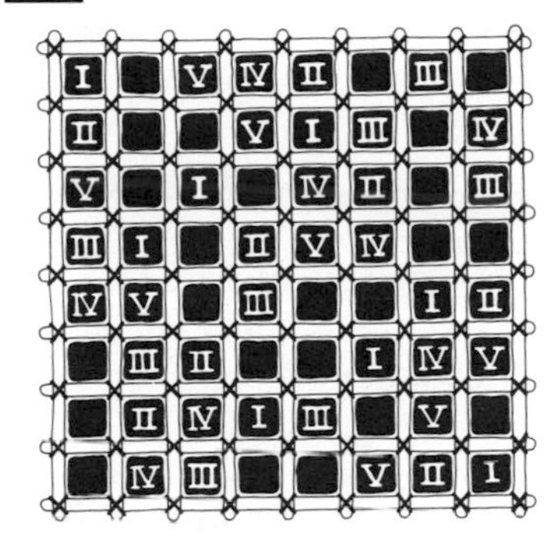

51 By opening the doors bearing the numbers 10, 6, 12, and 14, each row and column will total 20.

56 The numbers are thirteen and fourteen:

$$13 \times 13 = 169$$
$$14 \times 14 = 196$$
$$169 + 196 = 365$$

63

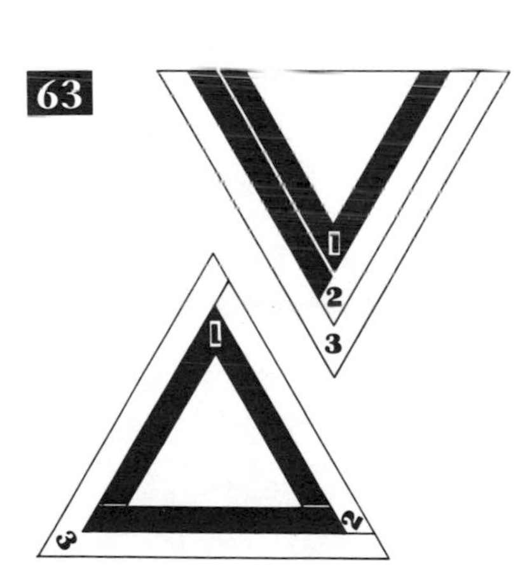

68 Since $7/8$ of an inch also represents the height of an equilateral triangle, 1 inch is the approximate base of this triangle. Hence, the total length of the strip is approximately 18 inches long.

75 The direction in which a "V" slot points determines the sunspot or sunspots in which a number must be totalled. The example below illustrates that the number in the "V" slot marked with an arrow must be totalled in the two sunspots identified with asterisks.

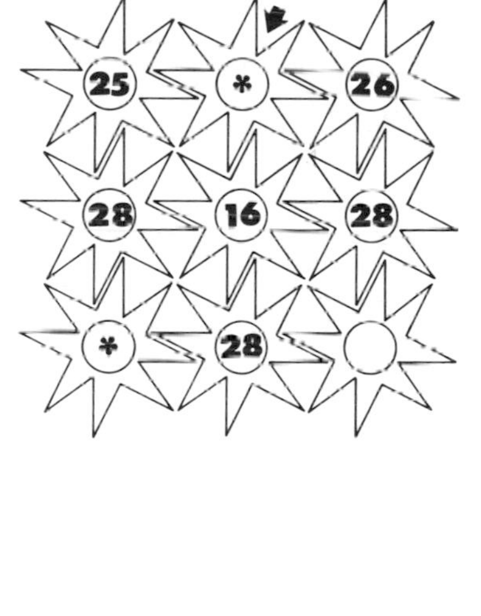

4

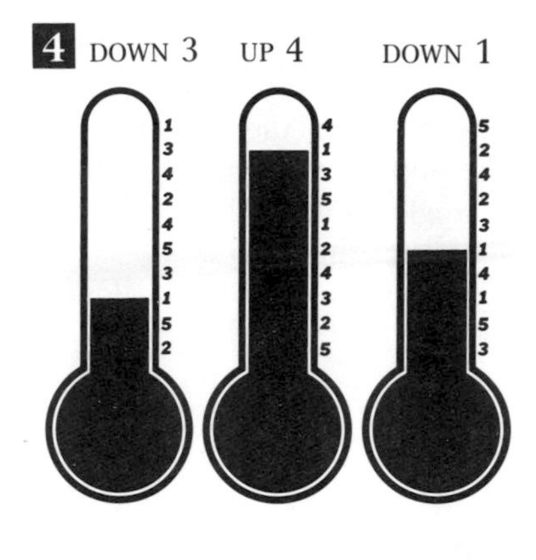

11

18

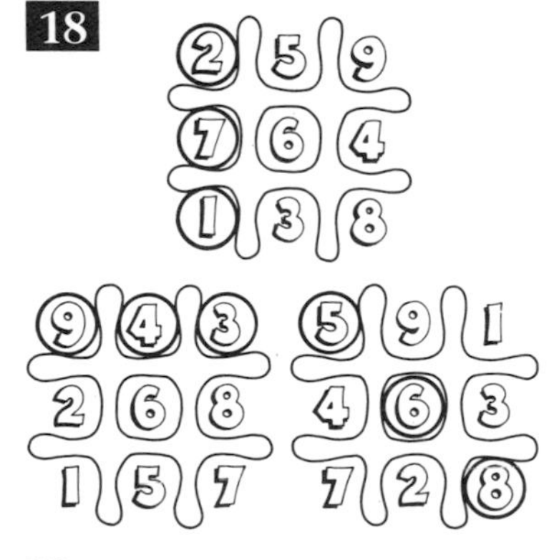

25 The total of all four comparisons is forty gallons. Since each color appears twice in the four comparisons, half this total is the total of all five colors. In the first two comparisons, red, green, blue, and yellow total eighteen gallons. Therefore, orange must total two gallons. Using similar comparisons, we find that red totals three gallons, blue totals four gallons, green totals five gallons, and yellow totals six gallons.

32

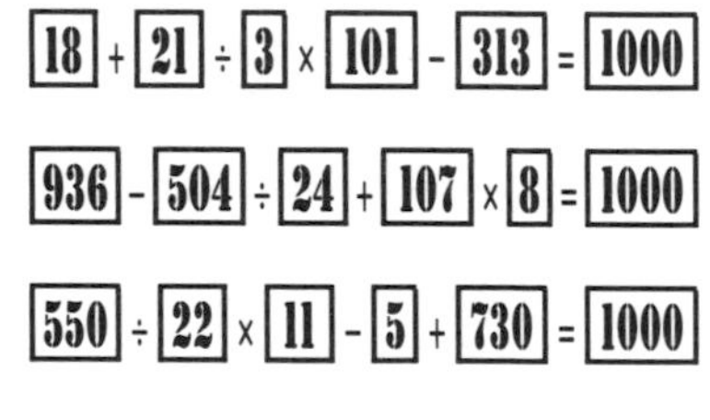

39

46

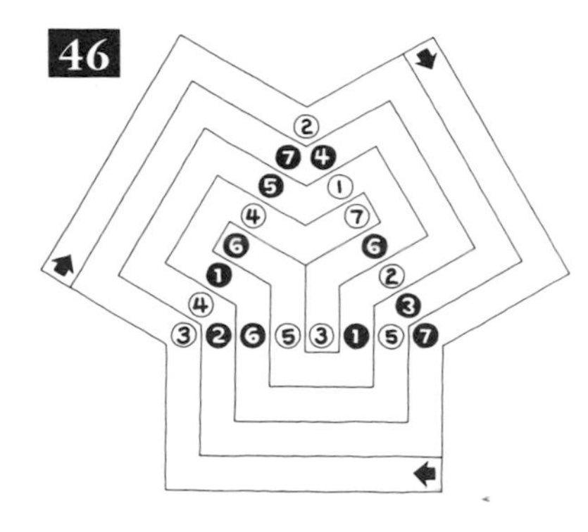

52

Mirror images also correct.

58

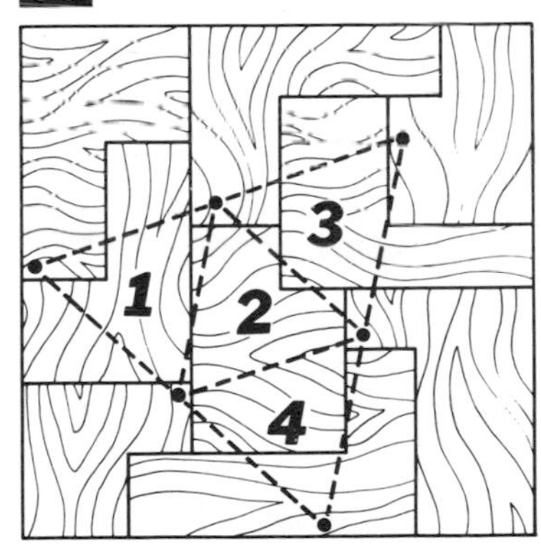

64

The remaining seven commands must be given in the following order:

Six squares right flank march.
Ten squares to the rear march.
Seven squares left flank march.
Four squares right flank march.
Thirteen squares to the rear march.
Fourteen squares left flank march.
Five squares left flank march.

70

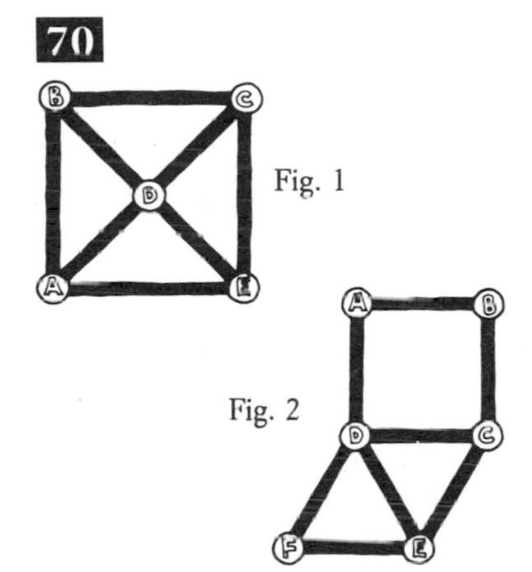

Fig. 1

Fig. 2

76

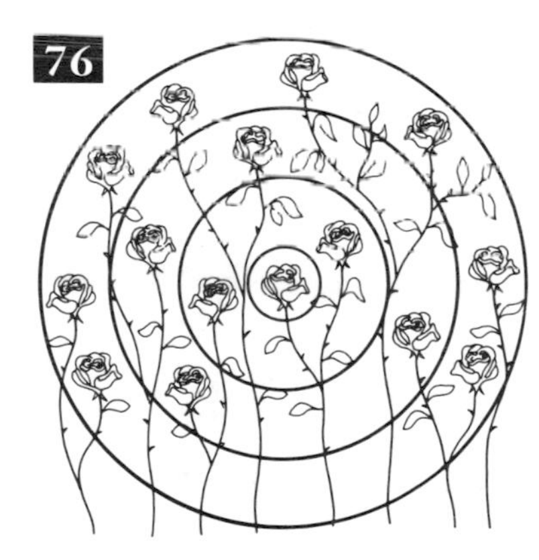

5

12

19

26

33

Sample on puzzle page:
1, -, -, 2, -, 3, 4, -, -, -, 5, -, -, 6, -, 7, 8.

Further solutions:
1) 1, 2, -, 3, -, -, 4, -, -, -, 5, 6, -, 7, -, -, 8.
2) 1, -, 2, -, -, 3, -, 4, -, -, -, 5, -, 6, 7, -, 8.
3) 1, -, 2, 3, -, 4, -, -, -, 5, -, 6, -, -, 7, -, 8.
4) 1, -, -, 2, -, -, 3, -, -, 4, -, -, -, 5, 6, 7, 8.
5) 1, 2, 3, 4, -, -, -, 5, -, -, 6, -, -, 7, -, -, 8.

40 Shown, twenty-one intersections. A total of 42 spots are required to create 210 intersections. That's one-quarter of the total spots times the maximum intersections per line ($1/4 \times 42 \times 20 = 210$).

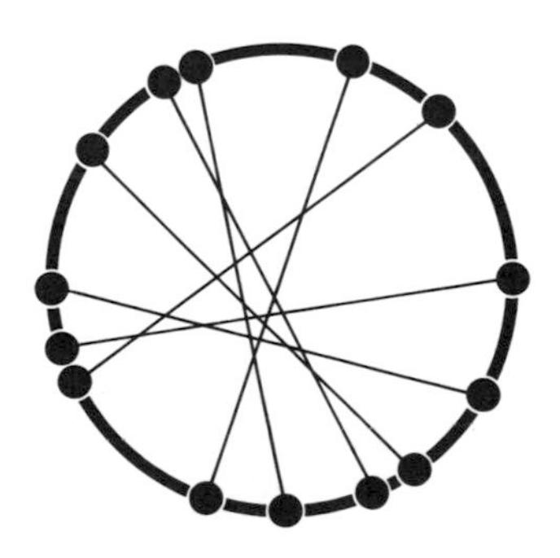

47

53 The top-right and bottom-left shapes were switched.

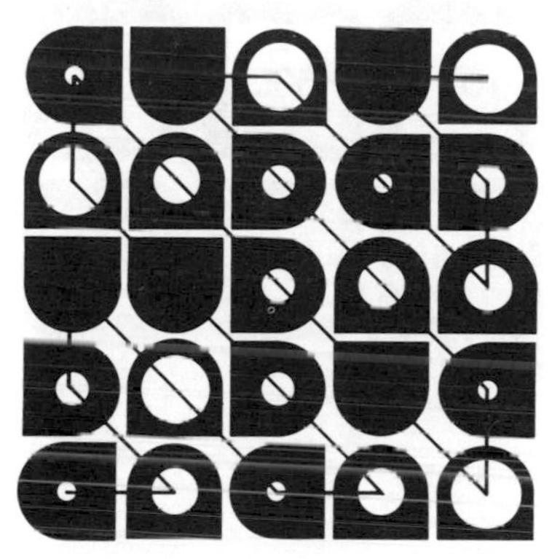

59

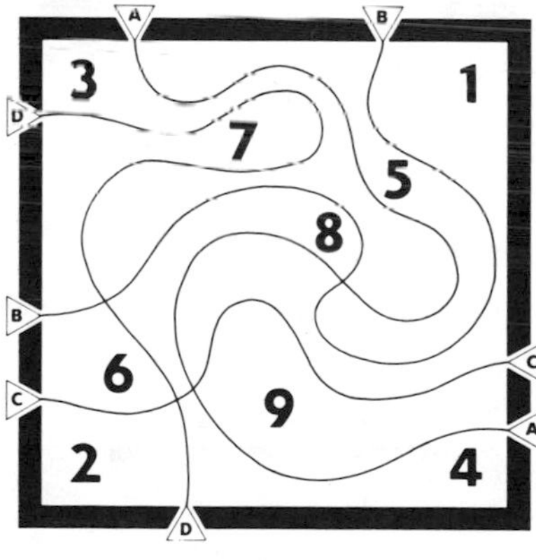

65 Numbers 3, 37, and 111.

71 As the gears appear from right to left:

233
▼
896
▼
594
▼
675
▼
483

77 Both craftsmen can solve in four pieces.

The glass cutter can solve with the fewest pieces, two, by flopping over either piece.

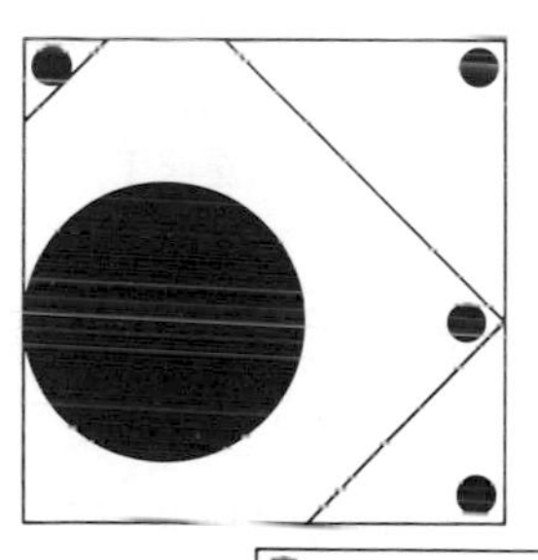

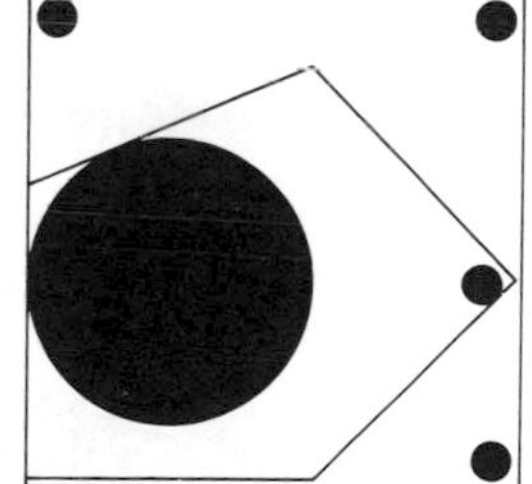

6

13 Cut four threads.

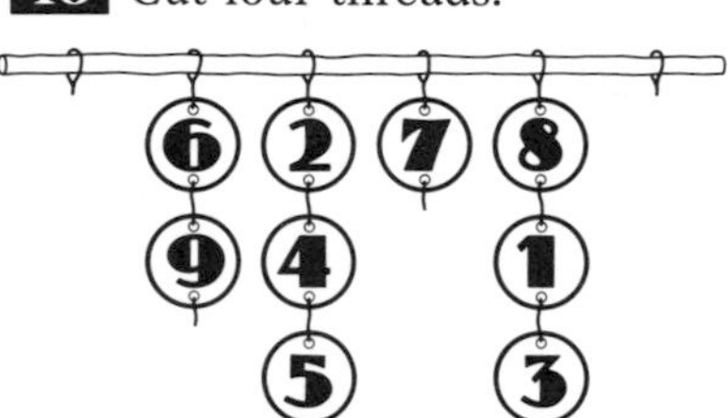

20 The lightning bolt equals the star. Using the letters L, M, and S to represent lightning, moon, and star, the following equations provide the proof:

given:	LMM = LLLSSSS
reduce to:	MM = LLSSSS
given:	LLLM = LLSSSS
therefore:	MM = LLLM
reduce to:	M = LLL
therefore:	LMM = LLLLLLL
therefore:	LLLSSSS = LLLLLLL
reduce to:	SSSS = LLLL
therefore:	S = L

27
Pie 1: $4\frac{38}{76}\% + 95\frac{1}{2}\% = 100\%$
Pie 2: $8\frac{27}{54}\% + 91\frac{3}{6}\% = 100\%$

34

602	709	204
107	505	903
806	301	408

41

1386

185
508

693

48

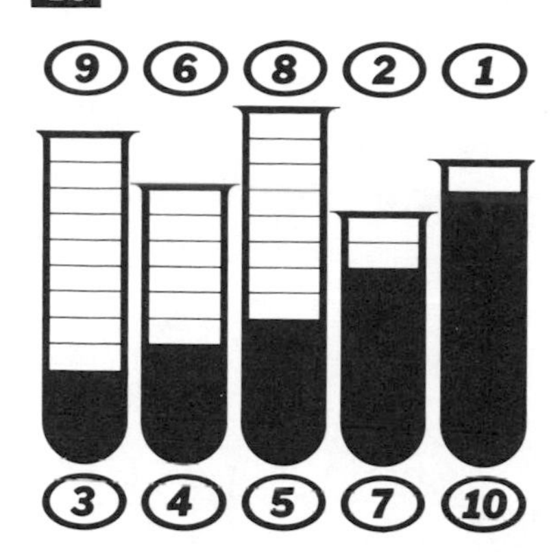

54 The correct sequence is: 3, 0, 4, 1, 5, 2, 6, 3, 7, 4, 8, 5, 9, 6, 10. Key: Subtract 3, add 4, subtract 3, add 4, etc.

60 The three numbers are one, two, and four. The divisions of each respective row are:

Row 1: 7, 7, 7
Row 2: 12, 6, 3
Row 3: 16, 4, 1

66

72

78

6 triangles + 3 = 9
12 triangles − 3 = 9
3 triangles × 3 = 9
27 triangles ÷ 3 = 9

7 This puzzle is solved by turning over, or "capsizing," the 9/8 hat. Now, 8/6 + 10/6 = 3.

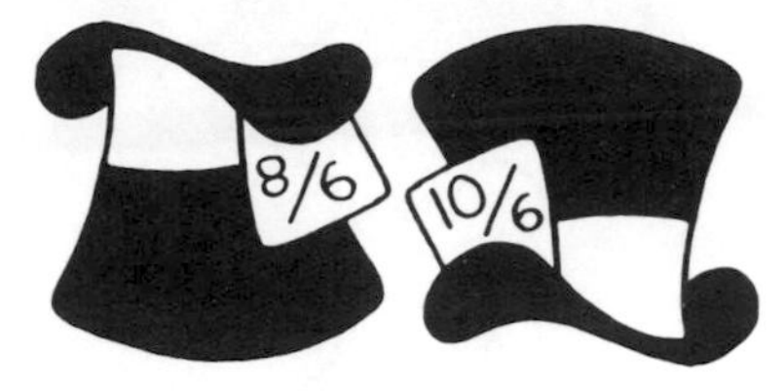

14

21

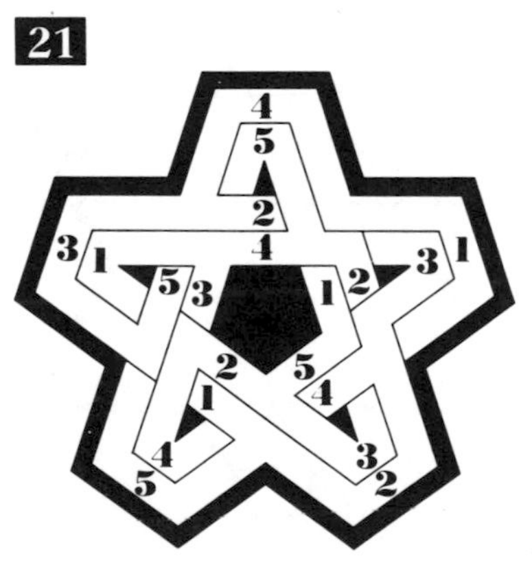

28 A stride of 52½ inches is reached on the forty-second step.

35

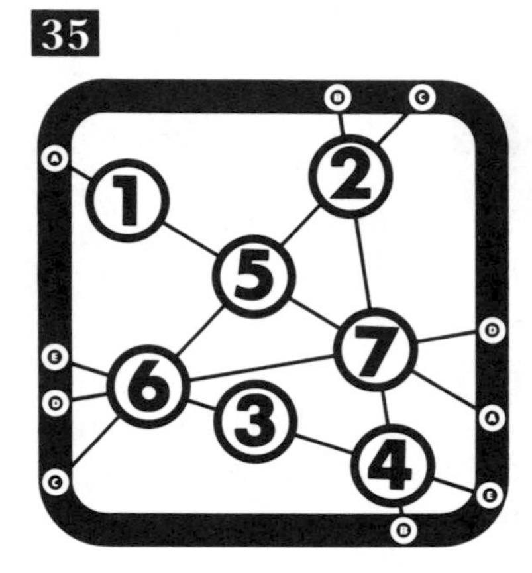

42

About the Author

If the world's greatest magician was Harry Houdini, and the world's greatest detective was Sherlock Holmes, then surely the world's greatest puzzle and game master is Steve Ryan, recognized as the most prolific creator of puzzles in the world with more than 11,000 brain-busting bafflers to his credit.

This virtuoso of vexation has been inventing games and puzzles since childhood. He eventually found a market for his creations through Copley News Service, where his *Puzzles & Posers* and *Zig-Zag* features have appeared for more than twenty years and currently challenge readers in more than 150 newspapers across the United States and Canada.

Ryan's creative genius has also catapulted him into television, where he co-created and developed the TV game show *Blockbusters* for television's most prestigious game show packager, Mark Goodson. Ryan has also written for *Password Plus*, *Trivia Trap*, *Body Language*, and *Catch Phrase*, and creates all the rebus puzzles for TV's *Classic Concentration*.

Taking inspiration from the world around him, Ryan's well of ideas never runs dry. As a firefighter for seven years in Southern California, he not only tested his cranium-crackers on the crew, but made puzzles out of nearly everything in the station. Fire hoses found their way into a maze, and other scramblers, math teasers, word games, and mechanical mind-wrenchers evolved from Ryan's analytical view of daily life.

Ryan is the author of such popular books as *Test Your Word Play IQ*, *Test Your Puzzle IQ*, *Pencil Puzzlers*, *Challenging Pencil Puzzlers*, and *Classic Concentration*, and the co-author of *The Encyclopedia of TV Game Shows*, the most comprehensive book of its kind. His puzzles have also appeared in GAMES magazine.

Many predicted Ryan's gifts in art, design, and mathematics would lead to architecture. But as usual, Ryan had a surprise twist in store: he built a mental gymnastics empire instead.

Nothing puzzling about that.

Puzzle Index

Page key: **puzzle,** *clue,* solution